HOW TO PLAN, DESIGN AND IMPLEMENT A BAD SYSTEM

How To Plan, Design and Implement A Bad System

Ronald B. Smith

Printed in United States of America

1 2 3 4 5 6 7 8 9 10

Library of Congress Cataloging in Publication Data

Smith, Ronald B

How to plan, design, and implement a bad system.

Includes index.

1. System design. I. Title.

QA76.9.S88S64 003 81-425

ISBN 0-89433-148-5 AACR1

When you fail to plan, you are planning to fail.

Dr. Robert Schuller

CONTENTS

PREFACE

The computer field abounds with controversy and different points of view. As Mark Twain said, "A man with a new idea is considered a crank until the idea succeeds." The book offers useful and practical ideas and approaches that have been tested on how to develop good application systems. For easy reading, as little computerese as possible has been used; however, the book is intended for executives and managers as well as systems analysts, programmers and EDP auditors.

The book has two related purposes. The first is to find realistic ways to eliminate the negative effects of developing systems. The second purpose is to look at approaches and alternatives to the traditional techniques of application development that can better serve and manage your organization.

Part I is dedicated to case studies of projects which used the traditional techniques of application development and went wrong. Included in the case studies are lessons that can be learned from this experience. Part II deals with alternatives to conventional application development techniques which include using software techniques. The last chapter concerns the integrated electronic office of the future which is available now.

Many of the case studies in this book are completely factual, although names and places have been changed for obvious reasons. I do wish to thank the publisher, O. R. Petrocelli, and my family whose efforts, support and understanding made this book possible.

I would like to express a very special appreciation to Mona R. Bertrand for her efforts in typing most of this manuscript and for authoring the word processing sections found in Chapter 11.

Ronald B. Smith

PART I

CREATING A BAD SYSTEM

1

FLOPS Flops

1

FLOPS FLOPS

Once there was a system to be designed called Field Labor Operational Payroll System (FLOPS). FLOPS started out as a good project, but ended up a dud and is still hobbling along. The purpose here is to show how things can go wrong on a project and the lessons learned from it.

There was a division that manufactured copying machines which was part of a conglomerate. The division's headquarters was located next to the conglomerate's headquarters. This division had branch offices in all of the major cities in North America. Each branch office had a terminal that was part of the conglomerate's communication network, which was hooked up to the main computer.

Each branch hired maintenance people from the local union to service the copying machines sold to customers. Each union had different pay rates and fringes for each job classification. The scope of FLOPS was to transmit the mechanics' weekly work data from the branch terminals through the conglomerate's communication network so that the mechanics could get paid. It was also to produce labor distribution reports.

NOTE: This chapter is based on the article "FLOPS Flops . . . But Why?" by Ronald B. Smith. Reprinted by permission of *Data Management,* July 1979.

PROJECT LOG

The following is the sequence of events during the systems development process:

January (first year). The feasibility study began, and a user project manager, a user project leader and two part-time analysts from DP were assigned to the project.

May. The users approved the recommendations from the feasibility study, and systems design began. The cost of the entire project was projected at $50,000.

October. The division elected a new president.

November. The analysts gave a systems design presentation to the users. The president rejected it because the turnaround time of the communications network was too long. In a hold position, all work on the project ceased.

March (second year). The systems design began again with a new cast, including a user project manager (president), a user project leader and a full-time analyst from DP.

April. DP upped the project price tag to $100,000 and a completion date of December. Also, DP decided (for the sake of saving time) to have programming concurrent with the systems design.

May–September. The following events happened during the redesigning and programming phase:

The user project manager (president) was too busy with normal duties to be actively involved with the project.

In effect, the new user project leader was an interface between users at the local branches and the analyst. The user project leader spent more time designing the system than gathering the requirements. Also, the requirements were forever being changed. Many of the controls the analyst wanted were diluted or deleted by the user project leader.

For unknown reasons, the programming manager disliked the analyst. This created a difficult internal work environment for the analyst.

The scope of the project was being enlarged.

Because of the above events and the fact that the analyst was trying to meet the implementation date of December, the analyst's intrapersonal relations became unruly.

October. Systems design was completed on schedule.

November. Final programming estimates were made. For the second time the cost of the project was raised, from $100,000 to $250,000. Also, the implementation dates moved from December to June. The users became very unhappy and, as an aftermath, the analyst was replaced with another analyst from DP.

February (third year). Testing of the system began, with many problems encountcred.

July. Testing and parallel runs were completed and the system was implemented.

August–December. The following events took place:

The users were still modifying the system.

The internal auditors audited the system and were unhappy with the loose computer controls and the absence of noncomputer controls.

Top management wanted to know why the project was so costly, why it took 2½ years to implement the system, why the system was still being modified and why overall controls left a lot to be desired.

LESSONS LEARNED

The scope of the above project had prearranged boundaries. If boundaries are continually being pushed back, then the chances of success are minimal. Also, the project should have realistic implementation dates.

Users play key roles during the systems development process. To be effective, the user project manager has to have the time to be involved in the project. The user project leader cannot reverse roles—playing both the designer and the analyst gathering the user's requirements. As one can see, users can make the difference in having a good or a bad system implemented.

If the analyst who designs the system is working under a tight implementation schedule, it might be better if he or she were the test analyst because of familiarity with the system. Also, this could shorten the testing phase. In other instances, especially from an auditing standpoint, it would be more advantageous if the system designer and the test analyst were different people; for example, the user could be the test analyst.

The internal auditing department should be involved in all phases of the systems development process. This should help educate and impress upon users the importance of good controls.

A project will become unstable when there is a large turnover of users and analysts working on the system. The continuity, interaction and transfer between the phases of a project will usually not be compatible and duplication of effort increases.

Concurrent programming with systems design on large projects has its benefits and drawbacks. The main benefit, if everything works, is saving of time. The main drawback is that the analyst changes specifications on something already programmed.

Due to changes in the business environment, there will be changes in the original requirements of the project. Due to changes in the technological environment, there will be changes on how the project will be designed, for example, IMS data bases and minicomputers. The longer the duration of a project, the more unknowns will be run into the corresponding costs.

RECOMMENDATIONS AND CONCLUSIONS

DP should produce a systems development process (SDP) pamphlet for DP, management and current-future project users. SDP is a structural approach to building information systems; it is simply a sequence of rational and logical thoughts and actions used regularly in making decisions. SDP is an essential tool that can increase individual contributions in project planning and development and can ensure project success. The aim is to educate the user in that role in SDP, decision points and risks.

DP should provide a standard control form to be signed by the user if he or she wants changes to requirements during the SDP. In this way, the user and DP will have documentation on one of the reasons why a project might go over its original cost.

Definition Phase.

Initial investigation—identifies the problem or opportunity and determines the need for further study.

Preliminary systems study—provides an economic evaluation and first definition of the proposed system. The economic evaluation includes the "ball park" costs of designing and implementing the system along with the operational costs of the current and proposed system. The first definition describes all major features of the proposed system. The study includes scope, objectives and constraints, identifies, evaluates and recommends alternatives, and plans the next step.

Systems analysis—describes the proposed business system in user terms. It provides a business solution to satisfy the user's requirements and assures that a technical solution can be developed and plans the next step.

Development Phase.

System design—describes the proposed business system in technical terms that detail the logical system. It also develops plans for conversion design and user procedures. The conversion design builds a bridge from the current to the new system. The user procedures prepare the users to operate the new system. The next step is planned.

Programming—defines the programs and the modules within the programs, and translates each module definition into machine-executable code. Finally, the programs are tested individually and as a complete system.

Implementing Phase.

Testing and installation—tests all aspects of the business system to the satisfaction of user and data center in a "live" environment.

Project evaluation — reviews how well the system meets user business objectives and develops recommendations for upgrading SDP. This review should take place from one to six months after the system becomes operational.

On Big Projects

All projects costing $25,000 or more should utilize PERT/CPM planning techniques starting with the preliminary systems study. On such projects, more work and the corresponding costs should be spent up front during the preliminary systems study. If not, the big project could exceed budget and implementation dates and be a poorly designed system.

Just as planning of a project is at two levels, so is budgeting. A firm budget is developed and a commitment made for the next step or phase. A "ball park" estimate is made for the total project. As the project proceeds, the estimate range narrows. Management is not asked to commit resources for the complete project until a clear picture of the solution is available. A complete solution might be obvious after the first step, or it may take a number of steps within SDP.

ON THE HUMAN SIDE

Since it is the user's system, the user must be made to realize that he or she has to have complete responsibility for controls. All too often the user does not know all the controls in a system and does not even realize the responsibility.

In FLOPS the analyst could not impress upon the user project leader the importance of good controls being built into and around the system. Therefore, the analyst should not try to mediate or correct the situation. In short, it could solve nothing and create ill feelings.

Most working people have one supervisor. An analyst has two—DP management and users. Since the analyst represents the users, it is his or her responsibility to let management know all possible consequences. If DP management decides to take no action about the situation, then there is nothing else the analyst can do. Therefore, the analyst must not be put in the position of trying to serve many.

FLOPS was missing one ingredient that brings stability to a project. It is called loyalty. For example, the programming manager put personal differences with the analyst ahead of the project. Also, the analyst was replaced after the final programming estimates were made. In short, the users held the analyst responsible for the increased costs and implementation dates. No gratitude was ever shown to the analyst for all the hard work put into the project.

No one person or event brought much harm to FLOPS. But, as one can see, the combination of people, time and events can help implement a poor system.

2

SLIPS Slips

2

SLIPS SLIPS

The Oh-Boy-Toy Company was in the business of manufacturing toys. The company's motto was "our work is boy's play." The toy company had its own computer, systems department and a payroll-personnel computerized system for the people who worked in their three plants.

One day the systems analyst stopped in to see his friend who worked in the personnel department. The analyst observed that his friend was writing on ½″ × 3″ white strips of paper and attaching the strips to a large table-top paper holder. Curiosity got the best of the analyst so he asked his friend what he was doing. His friend replied that he had two manual systems going for the plant personnel. The first was a seniority list for call-back from layoffs. The second was a seniority list for call-backs to home department—those people who are currently working but have been bumped from their home department. The analyst told his friend that all the information that he wanted resided on the company's payroll-personnel data base and listings could be generated to replace the manual systems. The analyst's friend thought the computerized system was a better idea and requested the new system.

The analyst filled out a System Request form for his friend which provided the following data:

Project Name: Seniority List Informing Personnel System (SLIPS)

Project Description: To provide the following two lists:
a. Seniority list for call-back to home department
b. Seniority list for call-back from layoff
Both lists will be sorted by seniority date and grade code.

Benefits:
a. Save 1500 hours a year for personnel in handling the company's three plants.
b. Time is of the essence in recall situations.
c. All information needed is conveniently located on one list.
d. Going from a manual system to a mechanized system.

PROJECT LOG

First week. The analyst completed the program specifications. The following is a copy of the specifications:

SENIORITY LIST INFORMING PERSONNEL SYSTEM

General	This program produces a listing by seniority of people who have been bumped from their home department and/or on layoff.
Input	Payroll-personnel records and plant parameter card.
Output	Report 1—Seniority list for call-back to home department. Report 2—Seniority list for call-back from layoff.

Frequency As requested.

Logic Step 1—Read all records from payroll-personnel data base.

Step 2—Bypass the following records:

a. that plant code from input plant parameter card does not equal plant code in record
b. that does not have a pay code of '03' (hourly)
c. that does not have employee status code of '00' (bumped) or '06' (layoff)
d. if the '00' record has a home department number that is blank, zeros or has the same number as present department number

Step 3—For all records selected, make a sort record consisting of the following field names:

PRINTER HEADING	FIELD NAME	FIELD CODE
LAST NAME	LAST-NAME	002
FIRST NAME	FIRST-NAME	003
EMP. NO.	EMP-NO	001
CLOCK NO.	STA-CLOCK-NO	036
STREET	STREET	004
CITY	CITY	005
STATE	STATE	007
ZIP CD	ZIP-CODE	008
PHONE NO.	HOME-PHONE	009
HOME SHIFT	HOME-SHIFT	182
PRESENT SHIFT	SHIFT	052
AVAILABLE SHIFT	AVAIL-SHIFT	180

HOME DEPT.	HOME-DEPT	183
PRESENT DEPT.	DEPT-NO	025
SENIORITY DATE	SENIOR-DATE	027
LAYOFF DATE	STAT-EFFECT-DATE	034
JOB CODE	JOB-CODE	031
GRADE CODE	GRADE-CODE	019
ACCEPT GRADE CODE	ACCEPT-GRADE	181

CLASS KEY—1 DIGIT

Use CLASS KEY '1' (Report 1) for all selected records that have employee status code of '00'. Use CLASS KEY '2' (Report 2) for all selected records that have employee status code of '06'.

Step 4—Sort on grade code within seniority date and seniority date within CLASS KEY.

Step 5—Move required data to corresponding print lines. See printer layout chart.

Translate plant code from input parameter card to plant name on report as follows:

01	Plant 1
02	Plant 2
03	Plant 3

An employee print set equals four lines—first three lines will be printed and the

> fourth line will always be skipped. On grade codes and CLASS KEY (Report 2) break, go to top of next page.
>
> The first page of Report 1 and Report 2 will be page 1. Page numbering is continuous for each report. Print 'LAST PAGE' at the end of each report.

Attached to the specifications were the input card layouts and the output printer layout charts.

Second week. The analyst turned the specifications over to a programmer for programming.

Fourth week. Programming is completed.

Sixth week. The system is successfully tested and turned over to operations for production. The analyst informed his personnel friend that the system was implemented and that the reports could be requested at any time.

Three months later. The analyst happened to walk by the doorway of personnel and noticed that his friend was writing on the same ½″ × 3″ white strips of paper and attaching the strips to the same large table-top paper holder. The analyst was shocked—he could not understand why personnel was not using the system he designed!

LESSONS LEARNED

This is not a story about how things can go wrong with a project. From a technical standpoint there was nothing wrong with the system. So why was personnel not using the new system?

There were two reasons. The first was that the new system

saved so much time over the manual system that the personnel people were afraid that some people would lose their jobs. This would never have happened because the company was growing so fast and everyone was needed. However, by using the new system it was possible that the company would hire one less personnel person in the future. The second and more important reason was that the analyst never made top management aware of the proposed system. This includes selling management on the benefits of using the system. You can have the best ideas or system in the world, but if you cannot convince top management, it is worthless.

The following are techniques that the analyst should use in selling his or her ideas to management:

The Human Element. Do not be a DP "know it all." Inject hypothetical situations and other alternatives or approaches. Relate to management in their terms.

Showmanship. Your presentation should include exhibits, slides, foils and a blackboard. A good habit to get in is to use a long pointer when walking through your exhibits, slides and foils. Be enthusiastic about your presentation and "trigger" ideas and questions.

Practical persuasion. Develop a simple systems presentation. Also, put yourself on management's level of speaking by explaining technical aspects in layman's terms.

Practical commitment. Do not oversimplify or oversell the proposed system. Sell only the truth! Also, deliver on time and within budget.

Closing. After your presentation, get an agreement on the next step of action before anyone leaves the room. The next step will include a specific time frame for completion.

3

The MAGIC Show

3

THE MAGIC SHOW

The Wizard Company was in the business of manufacturing magical toys in Wand, Georgia. Manufacturing management realized they had many problems within their own plant so they assigned two of their people on a part-time basis to study the plant problems and come up with recommendations. The user project team called their study Manufacturing and General Inventory Control or MAGIC. MAGIC was meant to be an ongoing and open-ended project.

The first area that the MAGIC study team tackled was Receiving. They recommended a complete reorganization, including installation of a computerized system. On an independent audit, Corporate Internal Audit gave the same recommendation. It was estimated that the company would save $10,000 a year in cash by computerizing the receiving area. See Table 3.1 for the cost/benefit summary.

The manufacturing study team did not know that their proposed receiving system had a direct relation to the accounting department's biggest problem—accounts payable backlog.

Table 3.1 Cost Benefit Summary

		Yearly Costs
1.	Hard Dollar Savings:	
	a. Replacement of data collection system	$ 5,000
	b. Eliminate need for leasing new copy machine and reduce paper requirements	5,000
	c. Labor savings resulting from replacement of present hourly clerk with a salaried clerk and reduced processing time per receival	15,000
	d. Saving expediting time for material control	5,000
	e. Elimination of receival packet from purchase order	5,000
	Total Savings	$35,000
2.	Proposed Receival System's Costs: Minicomputer, CRT, printer, communications adapter, modem and software	25,000
	Total Net Savings Per Year	$10,000

ACCOUNTS PAYABLE BACKLOG

This was a complex problem that did not originate in Accounting but ended up there. Accounting was on a manual matching system whereby invoice, purchase order and receiving report must match before payment. The core of the problem was when Accounting received the invoice but not the purchase order and/or receiving report. To break this down further, Accounting either (a) received the invoice with receiving report and no purchase order or (b) received the invoice with no receiving report.

Accounting did a study on category (b) for fiscal 1978. They averaged 331 invoice exceptions out of 2200 received per month. Therefore, the monthly percentage of invoices received that could not be matched to the receiving reports was 15%. The 15% exceptions did not include other problems from category (a) and approvals of purchase prices. The 15% plus the other problems came to a monthly exception rate of 24%.

A 1–5% exception rate is a level that Accounting could operate within. When the exceptions grew beyond 20%, then they became the standard and reached a level where they were too time consuming to be handled by the Accounting people.

Many of these exceptions were approved for payment by Receiving without any supporting documentation (packing slips and/or receiving reports). This practice violated corporate and all sound accounting practices and should not continue.

If the overall discount rate for fiscal 1977 was applied for fiscal 1978 purchases, then it was estimated that the company lost $12,000 in discounts not taken for 1978. In addition, the company lost $5600 for 1978 on discounts taken and later billed back. Part or all of the costs to rectify this situation could be offset by the savings of discounts lost.

The following changes could be the solution for the accounting department:

1. Use of packing slips as the receiving reports
2. Processing of all purchase orders no later than the end of each week
3. A complete reorganization of the receiving area which could include a computerized system

The above changes will not entirely correct the situation, but they will be a giant leap in that direction. Also, it could bring down the exception rate to a workable 1–5%.

PROPOSED RECEIVING SYSTEM

Step 1: Material arrives on receiving dock. The purchase order number is keyed into the video display tube for a match on the daily open purchase order file.

Step 2: If there is a match on purchase order number, go to Step 3.

Otherwise key in vendor number for a match on the file. If there is a match, go to Step 3.

Otherwise key in part number for a match on the file. If there is a match, go to Step 3.

Otherwise key in description for a match on the file. If there is a match, go to Step 3.

Otherwise hold the material in a no order area.

Step 3: Key in the receival data on the data display screen to update the file.

Step 4: List the 2-part Receiving Report on the attached printer. The crib and inspection areas each receive a copy of the report.

Step 5: At the end of the day, print the 4-part Daily Receival Summary Report. Accounting, Traffic, Expediter and Material Control each receive a copy of the report.

Step 6: At the end of the day, update the central computer from the local file.

Step 7: At the beginning of each day, update the open purchase order file on the minicomputer from the central computer.

See Figure 3.1 for the proposed receiving system. On the surface it appears that the system will help Manufacturing and Accounting. The system is designed and implemented.

Figure 3.1. Proposed receiving system

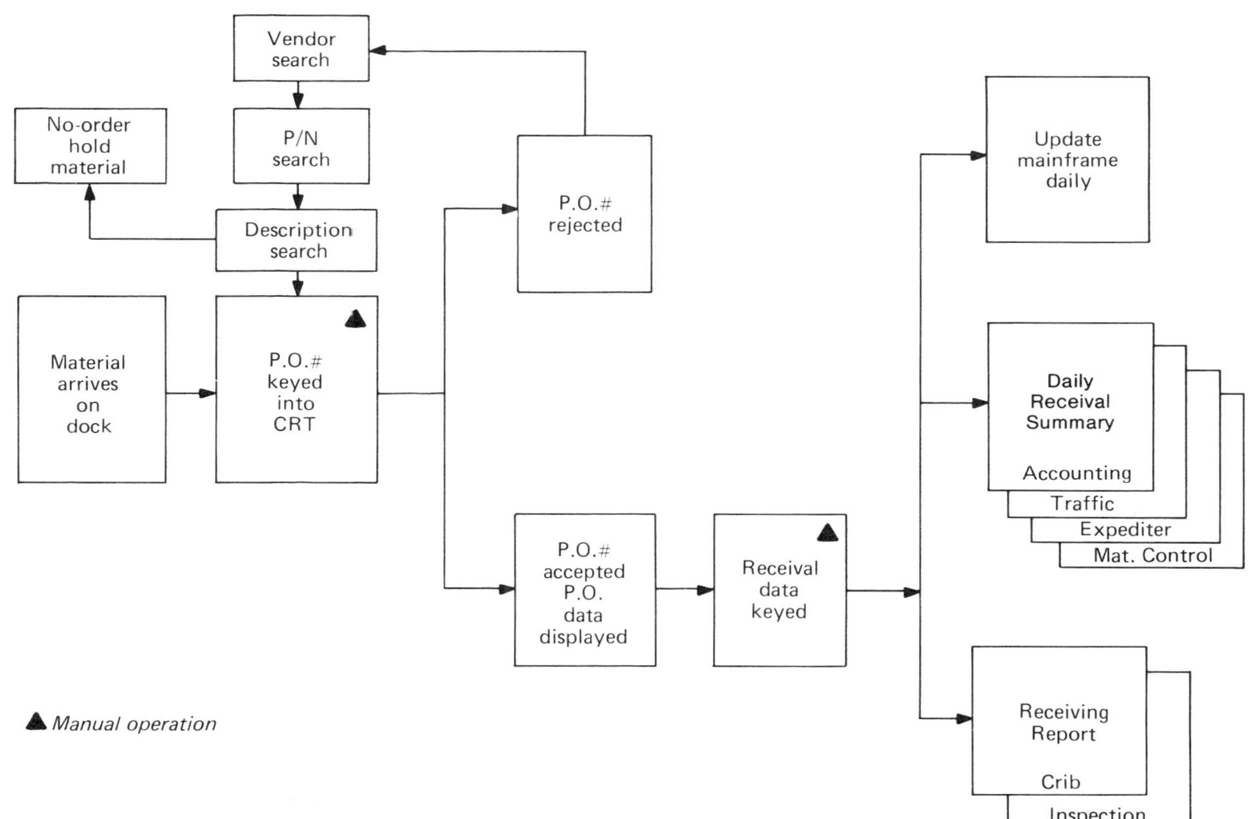

AFTERMATH

Below is a list of some of the problems with the system:

1. Too many open P.O. file searches. Also, one of the searches was on an alphanumeric description field which could lead to matching problems.
2. Embedded in the file searches are mental searches. For example, the vendor search might turn up ten P.O.s for the same vendor. What does the operator do now? The part number search might turn up ten P.O.s for the same part number. What does the operator do now?
3. No procedure or controls to handle the no-order hold material. In short, there is no follow-up system with checks and balances. For example, Receiving might receive material that does not belong to the Company and would not know it.
4. There are no monthly purges of the open P.O. file. For example, if there is no receival of the open P.O., it remains on the file forever. Eventually, the file will grow in size and could cause file space problems.
5. If the salary receiving clerk is late, sick or on vacation, who operates the system? Also, there was no user documentation on how to operate the system. Included in the procedures should be the controls to handle the no-order hold material.
6. As stated in the cost/benefit summary, the minicomputer was to replace the data collection system. In fact, Manufacturing kept the data collection system along with the new minicomputer.

CONCLUSIONS

It is very doubtful if the receiving system is cost justified. The manufacturing department did not get rid of the data collection system as stated in Table 3.1 which amounted to 50% of the yearly net savings. The minicomputer was not being used to its fullest capacity because there was only one application being run. Another application that could have gone on the minicomputer was an accounts payable system which would interface with the daily open purchase order file.

The manufacturing department should train another person to back up the salary receiving clerk in a time of need. Also, the manufacturing department should write up user documentation on how to operate the system.

Any system ever implemented for a department will always affect or interface with other departments in some shape or form. The manufacturing department did not know or did not care if their system interfaced with other departments. The receiving system escalated the accounts payable backlog problem. If Manufacturing had read about Accounting's problem and followed the partial solution of having all purchase orders processed no later than the end of each week, it would have eliminated most of the backlog problem. The time it takes to process the purchase orders, which in turn updates the open purchase order file, did make the difference in having an inefficient system.

The computer system should be reworked to eliminate some of the file searches and to have a new program to do a monthly purge of the open purchase order file.

4

FLAPS Flaws

4

FLAPS FLAWS

The First Line Automobile Parts Company of Syracuse, New York manufactured specialized auto parts for General Motors Corporation and American Motors Corporation of Detroit, Michigan.

The sales department had a manual system to compute and file sales quotes and margin markups for each part that the company produced. The sales of the company had doubled over the last three years and were expected to double again over the next three years. The sales manager, realizing that it was getting next to impossible to handle this growing manual parts system, asked the manager of the company's systems department to design and implement a computerized system to replace the manual system.

The systems manager assigned the best systems analyst to plan, design and help implement the system. The system was named after the company, First Line Automobile Parts System (FLAPS). The project, considered small in size and estimated to cost $14,000, was to be designed and implemented in 10 weeks. FLAPS required all input to the system to be generated via key-to-disk operations from the sales department. Computer processing would normally be on an overnight basis. The outputted reports would be delivered to the sales department on the following day.

The following are the three different key-to-disk operations:

1. *Maintaining the Index File by Index Transactions.* The index transactions create and maintain the parts on the index file. The index file is an index of parts which may be obsolete, replaced, seldom ordered or not found in the on-line part master file. The only conversion requirement of this system is to take the card deck which service parts personnel refer to as their "cookie jar" and convert this to the starting index file. The index file is one of the inputs into the next operation.
2. *Maintaining the Quotes File by Quote Transactions.* The quote transactions create and maintain the quotes on the quotes file. The purpose of this operation is to provide part number validation, item pricing and an 8½ x 11 quote (STD form) document suitable for mailing to customer. The effective days (life of quote) range will be from 1 to 120 days. Other than margin transactions, the quotes file is the only input into the next operation.
3. *Maintaining the Quotes File by Margin Transactions.* The margin transactions can maintain the quotes on the quotes file. The purpose of this operation is to give the sales department management the ability to select various quotations from the quotes file and analyze the margins by item and overall quotation.

See Figures 4.1 and 4.2 for the pictorial requirements of FLAPS.

Based on the scope and the requirements of FLAPS, the systems analyst designed the system and reviewed it with the sales manager. The review went smoothly and the sales manager agreed with everything.

Then the systems analyst turned over the systems specifications to the programming manager. The design was simple—three programs. The first program was to edit all the input data. The second program was to update the parts data bases with the valid input data. Finally, the last program was used to print the reports. In short, the systems analyst was applying the KISS (Keep It Simple Stupid) principle in the design of the system.

However, the programming manager disagreed with the design and decided to combine the edit and update programs into one program and assigned the job to a junior programmer. After 13 weeks the programming was completed and ready for testing.

Every time the edit/update program was tested it ABENDed. ABEND (abnormal end of task) is the termination of a task prior to its completion because of an error condition that cannot be resolved by recovery facilities while the task is being executed. The difficult thing was to determine if the records were being lost during the editing and/or updating process. It soon became apparent that combining the simple programs (editing and updating) into one program resulted in one big complicated program. After another four weeks of continuous testing, all the wrinkles of the system appeared to be straightened out.

The analyst wrote and turned over the user procedures to the users. Also, the users were trained to operate the data station. Finally, the system was implemented.

Figure 4.1 FLAPS flowchart of information flow

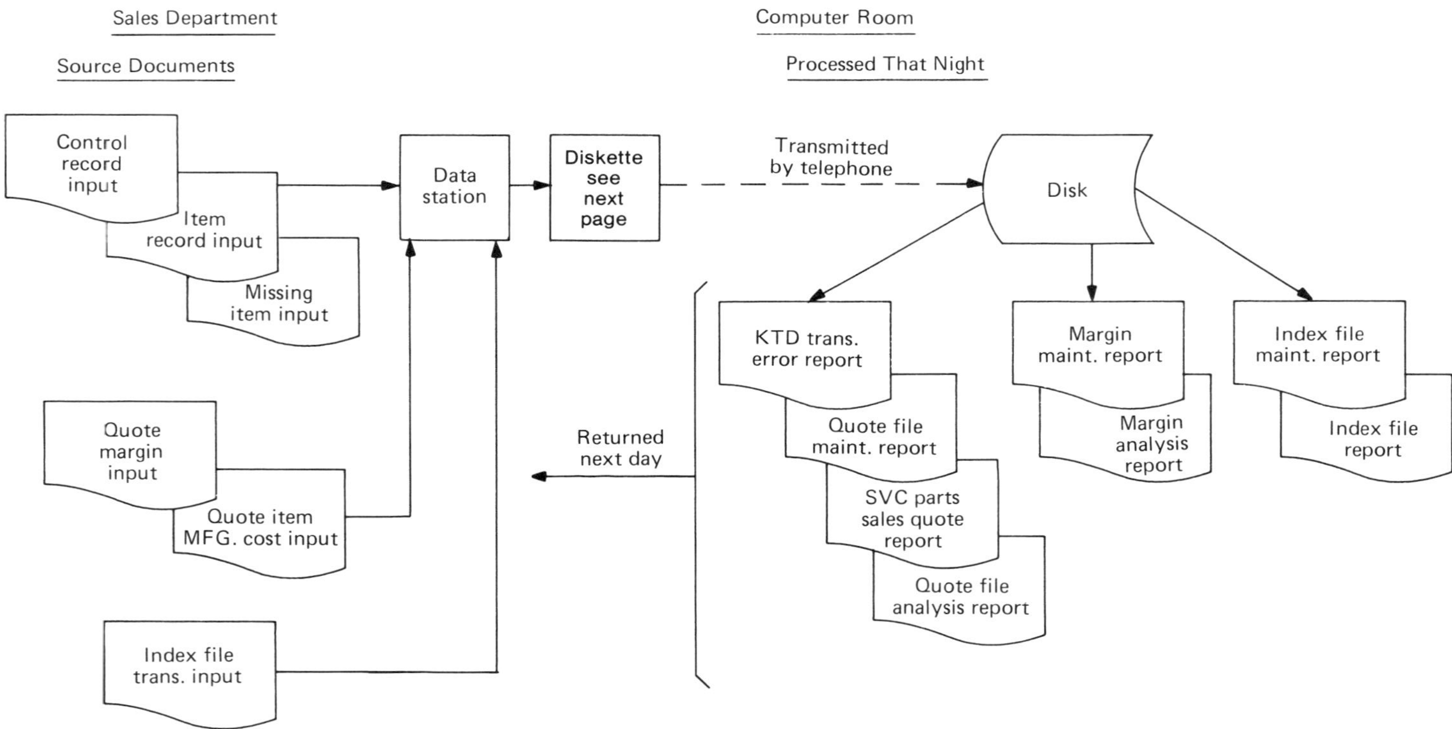

Figure 4.2 Overview of Diskettes

The entire system consists of 14 diskettes. On a normal day, one diskette from groups 1, 2 and 3 will be submitted leaving 3 backups within each group. On the upper left hand corner of each diskette is an external number label. The below boxes represent the diskettes with external number label detail and descriptions of contents:

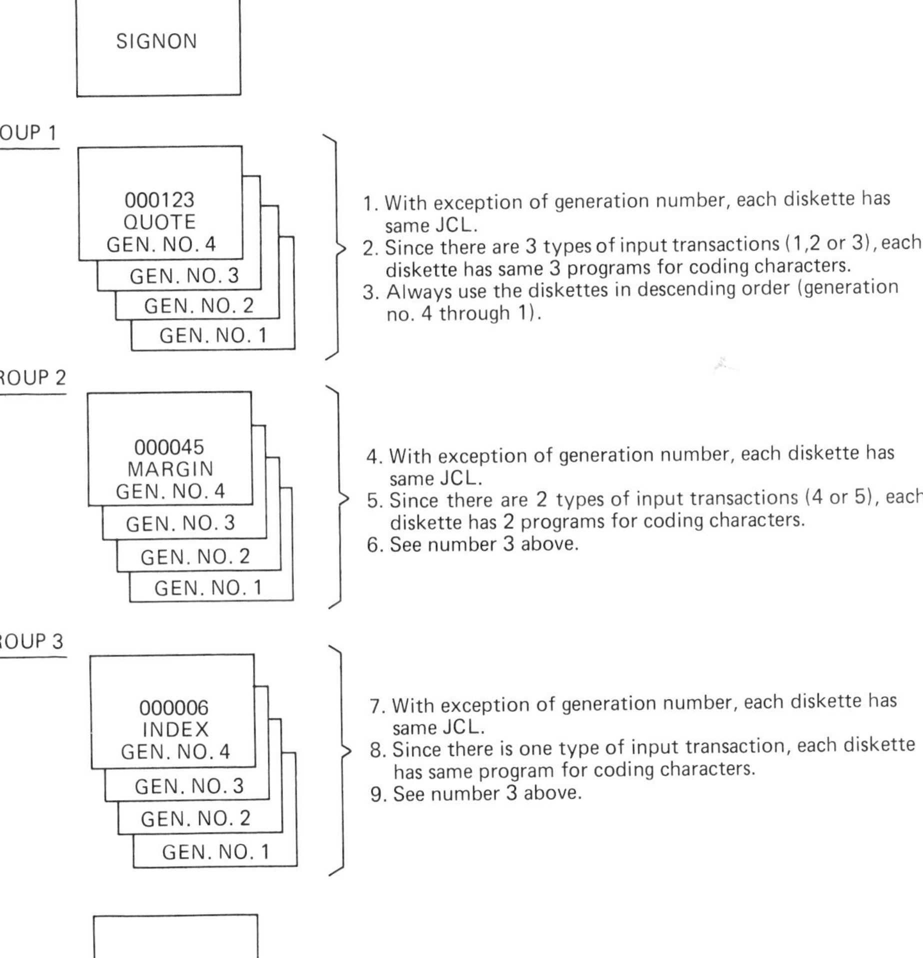

During the first month of operation, the junior programmer was called in on three different occasions (after midnight) because of an ABEND—records were still being lost within the edit/update program. After making programming corrections during the month, the system had no more ABENDS.

LESSONS LEARNED

Because of new programming techniques, there has been a recent trend toward cutting down the number of programs in a system by combining many functions into a single program. More recent experience shows that this is not always a valid assumption. In some systems combining numerous functions will work, while in others it will not work very efficiently. Each new system has to be looked at individually to find out if functions should be combined.

Figure 4.3 shows the analyst's systems design. The system should have cost $14,000 for 10 weeks of work with yearly production costs of $1,000. In addition, the system should have had a high program efficiency rating. Figure 4.4 exhibits the actual programming design. The system cost $25,000 for 17 weeks of work with yearly production costs of $1500. In actuality, the system has a low program efficiency rating. By comparing the two figures, one can readily see the differences. The system cost the users an extra $11,000 to develop with an extra $500 for yearly production costs.

Table 4.1 shows the 8 kinds of process combinations that can be found within any system. The chart rates the process combinations from the best situation to the worst. FLAPS falls toward the bottom of the range. The programming manager would have been better off giving the programming assignment to a more experienced programmer. Because of the lack of experience, it was not a healthy situation to put the junior programmer in. The junior programmer performed the best that he or she could do under the circumstances.

Sometimes it is human nature to take a simple solution to a problem and complicate it. Quite often, today's complicated solution creates tomorrow's problems. The lesson to be learned from FLAPS is that if you are fortunate enough to have a simple solution to a problem—keep it that way!

Figure 4.3 Systems design

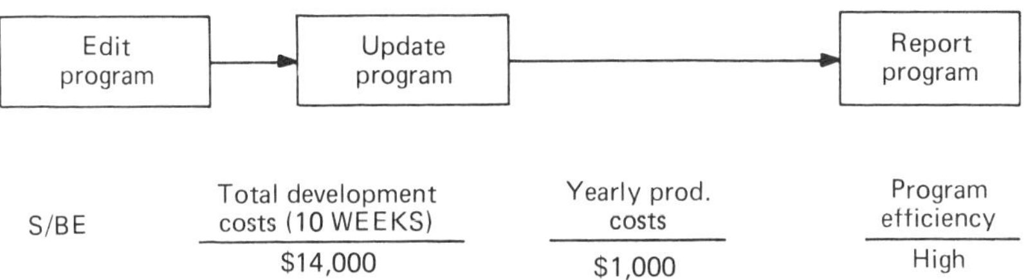

Figure 4.4 Programming

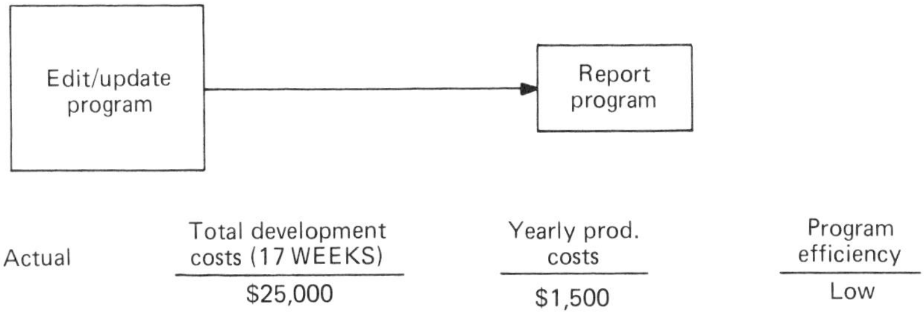

Table 4.1 Process Combinations

	Total development costs	*Yearly production costs*	*Program efficiency*
Best	Low	Low	High
	Low	Low	Low
	Low	High	High
	Low	High	Low
	High	Low	High
	High	Low	Low
	High	High	High
Worst	High	High	Low

5

The Pollution Solution

5

THE POLLUTION SOLUTION

A large diversified company in Canada that manufactures refrigeration units and related products is composed of 20 divisions that specialize in certain marketing areas. One of the divisions is called COLDQUEEN. COLDQUEEN came into being in 1970 and comprises four reporting groups—engineering, manufacturing, service parts and accounting. This story is about the archaic service parts system, its problems and a possible solution.

HISTORY

The following is the history of the service parts system.

1970

The manual system consisted of shippers, invoices and balance-on-hand cards. The balance-on-hand cards were continually maintained for additions and deletions. The product lines were truck/trailer, bus/rail, container, marine, military and aircraft. The truck/trailer, bus/rail, container and aircraft parts were handled by separate individuals. The marine and military parts were handled by one person because it was a manual job-cost system.

1971

The manual system was broken down into functional work groups: order handling, material control and customer service; each group was headed by a supervisor. The six different product lines were lumped together for inventories, ordering, typing shippers and invoices. In summary, the people were cross-trained for all of the product lines.

1973

The service parts group realized it needed a mechanized system to replace the manual system. The corporate system's department suggested it copy the service parts computer system that a larger division was using. Therefore, COLDQUEEN copied and implemented most of the service programs from the larger division for a cost of $20,000.

The following were the key reports from the batch system.

Adjustments—back orders
customer address
invoices
part number chains

Back order status

Customer master records

Gross margin analysis

Invoices

Location of part numbers in warehouse

Open shippers created and not billed

Sales orders

Service parts year-end inventory

Stock status

1974

Service parts spent $12,000 to add a few programs and for maintenance of the system.

1975

Service parts spent $14,000 to add a few programs and for maintenance of the system.

1976

Service parts spent $16,000 to add a few programs and for maintenance of the system.

1977

Service parts spent $18,000 to add a few programs and for maintenance of the system.

1978

Service parts spent $20,000 to add a few programs and for maintenance of the system.

1979

Service parts spent $20,000 for maintenance of the system.

The above excludes the cost of running the system. So far, the service parts group has spent $120,000 for the system and this amount will grow every year. Spending more money on this system is a complete waste.

CONCLUSIONS

In 1973, the system that was copied was fifteen years old, had gone through three different hardware conversions and was saturated with maintenance changes. Also, the system was designed with a lack of programming standards. Therefore, this was a polluted and outdated system that COLDQUEEN was buying into.

The system did, however, have some benefits:

Eliminated two typists

Produced shipper within 24 hours of order

Generated back order shippers within 24 hours after material became available in inventory

Faster invoicing

More control over shipping/billing functions

The following were the short-term problems associated with the system:

No user training

Learning curve

Unfamiliar with output

Growing pains

Long-term problems with the system were extensive.

First, one mistake in entering data into the system has a ripple effect in the system by causing six or seven mistakes. To correct all the mistakes is a very time consuming venture which service parts people do not have the time to do. Some of the mistakes cannot be corrected or explained. This is called the full moon, raining and midnight in China syndrome.

Second, sometimes the service parts people had to substitute a manual transaction to correct an error by inputting a transaction into the system to force the system to have the right outcome. For example, service parts received an incorrect invoice and its total card. Service parts destroyed the two outputs, typed an invoice and keypunched a new card. To circumvent the system, the new total card was inputted into the total transaction program. Therefore, there was too much manual intervention in the system.

Third, part card system.

Fourth, insufficient controls to monitor the total system.

Fifth, no documentation of system and no user training procedures for new employees.

Sixth, lost output from the computer center caused by poor controls by the output section of operations.

Seventh, since the system was in batch mode, long delays were common between data preparation and the outputted reports. For example, the inventory reports may be out of date before they are finished printing.

Eighth, communications with operations were poor because of the following:

Service parts people worked the first shift.

Operations ran the system on the third shift.

Operations had a high turnover of people.

Service parts people were not always talking to the operational people doing the work. Therefore, no working relations were built with operations.

The more people involved caused more talking, less productive work and more chances for mistakes and/or misunderstandings.

When problems came up, no one knew who had the responsibility (user, operation, programming) to correct the situation. This type of situation often caused friction between groups and magnified problems. The system needed an ombudsman.

For all the money COLDQUEEN spent, they could have bought an on-line order entry and inventory control package with the following features:

1. On-line transaction processing and interactive inquiry to facilitate more accurate business decisions based on timely information.
2. Video-display terminals to maintain a file of customer names, service part descriptions and both open and shipped orders.
3. Display of the status, location and inventory of every part. As orders are entered, shipped or back ordered, inventory levels are updated immediately.
4. On-line editing to eliminate batch edit/error reports.
5. Override ability for all data items.
6. On-line functions which include the following:

   Order entry

   Order maintenance

   Order retrieval by open orders, hold orders or customer

   Customer master file maintenance

   Invoicing

   Stock status

   Transaction register

7. The batch part of the system which includes the following:

   Inventory

   Backlog

   Standard cost

   Gross margin analysis

   Audit trail

   Back order list

   Shipper and invoice documents

   Daily sales transaction analysis

   Daily shipment analysis

Some of the advantages of this package are listed below:

1. Job enrichment
2. The system can be learned quickly and new operators can be trained easily
3. Documentation
4. Implementation and technical support from the company that supplied the software package
5. Built-in controls
6. Extensive editing
7. Override ability
8. Input batch balancing
9. Little or no money spent on the maintenance of the system
10. Since all customer-related data is available through on-line retrieval, customer inquiries can be answered immediately, thereby improving customer relations

When a company needs a system, one of the first tasks should be software analysis and application analysis. This analysis should include reviewing computer program packages already in-house to determine if they can be used. If not, software packages sold by outside vendors should be evaluated for possible use. Going to an outside vendor can save a company much time and money.

6

The Weak Link

6

THE WEAK LINK

Accounting and data processing professionals have related that the documentation of controls for computer processing is weak at best. Much personal experience agrees with that premise.

How often does a DP auditor audit a computer system and find the following situations?

a. No documentation
b. Outdated documentation
c. Current documentation but not enough of it
d. Everyone's documentation on the system is different
e. Some key people have documentation and others do not
f. The user left the company and took the system's documentation to his or her next job
g. No sign-out and/or control procedures from the library within the systems department
h. No library in the systems department

NOTE: This chapter is based on the article "Model for Documenting Controls Strengthens Weak Auditing Link" by Ronald B. Smith. Reprinted by permission of *Data Management,* May 1977.

LACK OF DOCUMENTATION

The use of computer-based business systems brings with it special needs to establish and document controls over the entire processing system. Many times there are adequate controls in a system but the documentation is not standardized. The general purpose of controls in any system is to ensure that all data is entered accurately into the system, processed correctly and the results verified.

The purpose of this chapter is to illustrate a model for documenting controls for computer processing. This model can be used by the systems designers, user department management, internal auditors and independent auditors. When writing the controls for this model, it is best to use as little "computerese" as possible. In this way, the nontechnical reader can better understand the control contents of the system. If the auditor wants more data, he or she can always go to the information processing department for more backup details.

The following list reflects the eight standard control sections of this process model:

Control section	*Title*
A	Controls Overview
B	General Narrative
C	Input Preparation
D	Input Transmission
E	Data Preparation
F	Computer Processing
G	Output Processing
H	Output Reconciliation

Case in Point

Payment Entry processing will be depicted in this model.

Control Section A—Controls Overview as indicated in Figure 6.1. The pictorial overview should come first because it is much easier to understand an illustration of the flow of processing than just reading words about the processing. For continuity in all overviews, it is best to use the same standard symbols. Always include the user's title or user's department on the overview; this can be especially helpful to the user department management. The key (control section and title) is indicated in the overview for easy reference.

Figure 6.1 Controls Overview

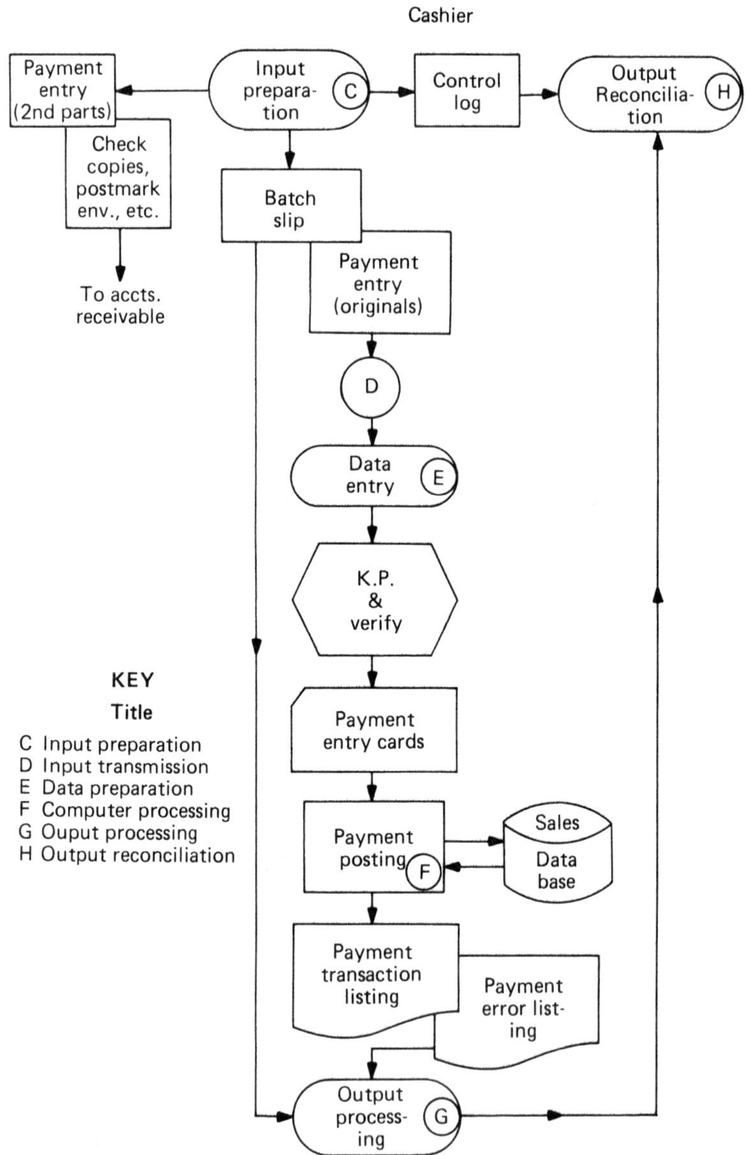

Control Section B—General Narrative as indicated in Figure 6.2. This is a brief description of what is taking place in payment entry processing. This should come immediately after the pictorial overview.

Figure 6.2. General narrative

CONTROL SECTION B—General Narrative

Payment entry is the process by which payment for finished goods that are shipped and invoiced are registered on the data base. This posting reduces the customer's total indebtedness figure and creates a payment record for use in later reconciliation to accounts receivable detail records.

It is essential that this function be performed rapidly and with extreme accuracy.

The methods used to control this processing are detailed in the other exhibits.

Control Section C—Input Preparation is indicated in Figure 6.3. From this control section on, parts of the pictorial overview are to be shown again, relating them to a corresponding control section. This saves time because it is not necessary to refer to the pictorial overview. The partial overview should be the same size and dimensions as it is depicted on the pictorial overview. Also, each partial overview will be indexed by its control features. Below the partial overview, the applicable control features will be described. The input preparation controls are the prenumbered forms, batch slip and control log.

Figure 6.3 Input Preparation

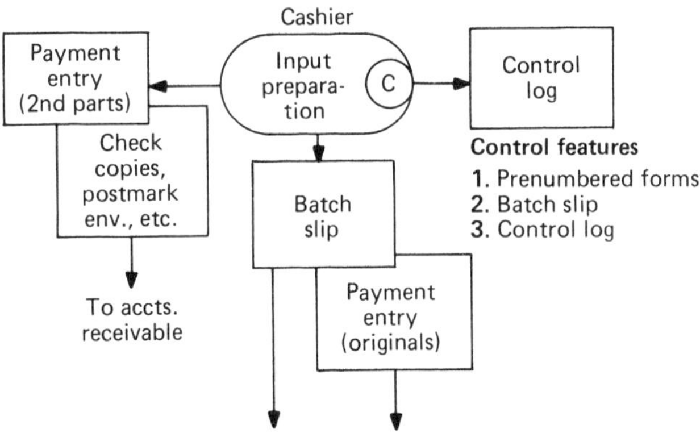

Control Section D—Input Transmission as indicated in Figure 6.4. From this exhibit, we can see that there are no controls. This could be an indication that there is a lack of and/or need for controls. If there is a need for controls, the situation should be investigated and, if possible, corrected.

Figure 6.4 Input Transmission

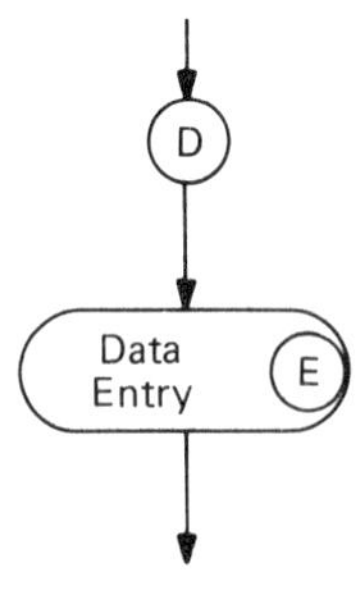

Control Section E—Data Preparation as indicated in Figure 6.5. The data preparation controls are batch checking, key verification and inclusion of batch number on each input record.

Figure 6.5 Data Preparation

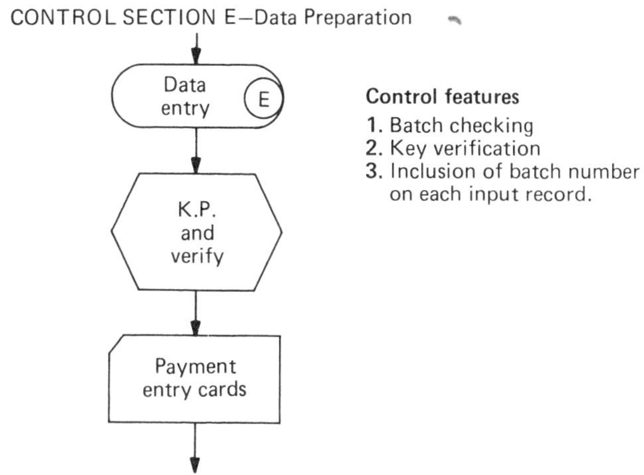

Control Section F—Computer Processing as indicated in Figure 6.6. The computer processing controls are input validity tests, error log output, transaction listing with control totals and input-output total control.

Figure 6.6 Computer Processing

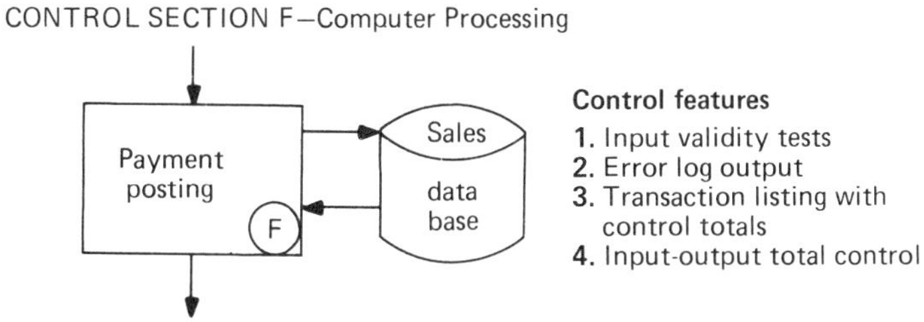

CONTROL SECTION F—Computer Processing

Payment posting

F

Sales data base

Control features

1. Input validity tests
2. Error log output
3. Transaction listing with control totals
4. Input-output total control

Control Section G—Output Processing as indicated in Figure 6.7. The output processing controls are batch numbers on output.

Figure 6.7 Output Processing

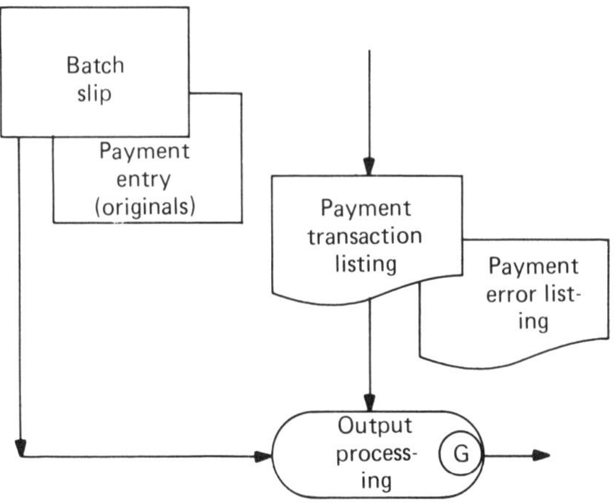

Control Section H—Output Reconciliation as indicated in Figure 6.8. The output reconciliation controls are a comparison of control log to output reports and a comparison of returned input batches to control log.

Figure 6.8 Output Reconciliation

A Standard Procedure and Standard Practice are referenced in this section. A procedure is a document that tells people how to proceed to do work. It tells a number of people or a number of groups how they fit into a single team play. The procedure instructions are written in a playscript format.

A practice is a singular version of a procedure. It is a document that tells one person how to proceed to do work. Therefore, the playscript format is not used.

Most Important Controls

The control features in this example cover a wide variety of techniques and methods. There are many control features that were not mentioned because they were not applicable to this case study.

In some cases a control section might be missing from control sections C through H because it is not needed or is not part of the particular processing. The two most important control sections are F, Computer Processing, and H, Output Reconciliation. These two sections should always be present and well documented. Also, in some cases a control section might be in a different sequence. For example, control section H, Output Reconciliation, might come before control section G, Output Processing.

The controls write-up can reference outside documentation. For example, Standard Procedure and Standard Practice were referenced in this case study.

Benefits of Controls

The user department management will take a more active part and interest in the system if it is involved in approving the controls of the system. Someday that part of management's job requirements will be knowledge that the system has adequate controls or helping implement the controls and responsibility for the controls. The user cannot always take for granted that the information processing department implements the proper controls for the system. Involvement is the key to educating the user and it gives the user more confidence in the system.

The controls of the system should become part of the systems documentation book that the information processing department prepares for the user. Also, the information processing department should retain a copy for its own documentation. Internal and external auditors always give high marks when they see this type of documentation on systems. Below are some reasons why the controls write-up is a good reference for the future:

As a reminder as to what the present controls are.

Training new employees.

Auditors who go directly to the information processing department can see their copy.

When modifying the system, the controls should be checked for possible updating to make sure none of the controls were lost or if new ones were added.

If the present system is to be replaced with a new system, the present controls could be used as a guideline for setting up the new system.

From a systems point of view, writing up and getting management's approval of the controls has its benefits for the systems analyst. After approval, the systems analyst can write up the specifications of the system for the programmer to code. Knowing exactly what the controls will be beforehand makes the job of writing the specifications much faster and easier for the systems analyst.

CONCLUSIONS

The controls write-up, Standard Procedures and Standard Practices, should be part of the systems documentation book. Other sections within the user manual could include:

User manual distribution list

Application narrative

Flowchart of the information flow

Output descriptions with sample reports

Input forms

Definition of error correction procedures

Interfaces with other systems

Miscellaneous and correspondence

As one can see, there are many intangible benefits of having good documentation that are difficult to quantify in dollars and cents. It is important to recognize that one of the biggest areas of savings is in future time since the cost of time in the future will be more expensive than now.

The need for good documentation is becoming a very essential element for a good internal control system. This process model is offered as one vehicle for internal control since it is flexible enough to be modified to suit a company's own requirements.

7

FLIPS Flops

7

FLIPS FLOPS

The Adventures-In-Wallpapering Company was in the business of manufacturing wallpaper and paints. Six months ago the company's systems department designed and implemented Factory Labor Information Payroll System (FLIPS) at a cost of $200,000. The system has been operating without any known problems.

Recently, the internal DP auditors audited the system and found a number of weak areas:

1. The field that was used to identify each factory worker was their social security number. When one thinks about it, social security numbers are not "unique" enough to identify individuals. For example, there have been cases where duplicated social security numbers have been issued to two or more people. Also, a factory worker could receive two or more paychecks by having at least two different social security numbers. To add to the confusion, there are different sets of brothers in the plant whose social security numbers differ only by the last digit.

2. The total regular hours inputted never equal the total regular hours paid.

3. The total overtime hours inputted never equal the total overtime hours paid.

4. There is no maximum range test for the hourly rate field. Theoretically, the hourly pay rate for new people or for pay raises could have been inputted as $99.99. The maximum

range test should have been set at between $15–20 per hour.

5. There is no maximum range test for the total number of hours worked per week. There should have been an Hourly Exception Report for people who worked more than forty or fifty hours per week.

6. The Input Error Report had listings of the input with its error message. The problem was that the total number of possible messages is too few and generalized. In short, most of the time it was difficult to determine exactly what caused the error.

7. The system ran continuously through the first program to the last program without any stop-and-go decisions being made during the run. For example, if the total calculated hours were off 5% or more from the total hours inputted, there should have been a decision made not to run the rest of the system.

8. If a factory worker left the company, he or she would remain on the file forever. There should have been a once a year program in January (after the W-2 forms are printed) to write out these records to a past employee file and then delete the records from the active payroll data base.

9. The weekly production costs of the system is 50% greater than stated on the original cost-payback analysis.

After the post audit, the DP auditors asked the systems department how much it would cost to make the programming changes to take care of the control deficiencies of FLIPS. The systems department costed out the modifications for a total price of $20,000.

The question now is whether the company wanted to spend another $20,000 on fixing the system.

LESSONS LEARNED

We are not going to try to answer the $20,000 question. In fact, at this point, it is almost immaterial. A DP auditor should have been involved in most of the phases within the systems development process (SDP). The purpose of this example is to show the involvement of the DP auditor in the most crucial phase of SDP. This phase is called systems design, which is basically the time when the working relations between the parts of a system in terms of their characteristic actions are specified.

In the past auditors have audited computer systems after they have been implemented. This is auditing after the fact. Often the cost is too prohibitive to go back and modify the system to incorporate adequate controls. Audit controls should therefore be built in during the system's design phase and not added after implementation of the system.

The DP auditor should be part of the systems development team which includes the users. His or her role within the team would be as an independent advisor. An independent advisor does not give or take orders from anyone on the team. The auditor's role is to make judgments on the controls and to make recommendations to the team. These are based on what might happen after the system is implemented. The auditor's evaluations and recommendations should be written up and copies given to management, members of the systems development team and the corporate auditing team.

Guidelines to Follow

The DP auditor might use the following guidelines in his or her evaluations and recommendations:

1. Define and recognize control objectives as separate design requirements.
2. Look at the following control areas with emphasis on audit controls, audit trails, error correction procedures and adequate design documentation:
   a. Input preparation by user
   b. Input transmission by user
   c. Data preparation by the DP department
   d. Computer processing
   e. Output processing
   f. Output reconciliation

   The two most important areas—computer processing and output reconciliation—should always be present and well documented.
3. Review the original cost/benefits analysis to find out if the design phase was within budgeted cost.
4. Reexamine the budget costs for the subsequent phases within SDP for any changes.

Auditor as Educator

The DP auditor and systems analyst should have a close working relationship. The systems analyst has two main goals. The first and primary goal is to design a system according to user requirements; the second is to have adequate controls in the system. The DP auditor's only goal is to have adequate controls in the system. Since the primary goals of the systems analyst and auditor differ, there has to be a meeting of the minds. It is up to the auditor to educate the systems analyst on the importance of adequate controls. In fairness to the systems analyst, he or she cannot always implement adequate controls because of user requirements and/or the need to meet tight implementation dates.

Since it is the user's system, he or she must be made to realize that the user has to have complete responsibility for the controls. All too often, the user does not know all the controls in the system and does not even realize that he or she should be responsible for the controls. The user should know that the system must have the proper controls and should be aware and be responsible for all of the controls. This has to be part of the user's job description.

Role of Top Management

To make the role of the DP auditor during the systems design phase and the other phases within SDP a success, top management must mandate certain responsibilities for the user, DP department and DP auditor.

Users must be aware and responsible for all controls in their present systems, ensure proper controls are implemented in proposed or new systems and ensure any modifications to the current systems have adequate controls.

The DP department must include the DP auditor as part of the systems development team and, as much as possible, cooperate with him or her. The DP auditor should also be made aware of any design modifications to current systems and be invited to any state-of-the-art training sessions for personnel.

The DP auditor must have sufficient DP training and knowledge to make the above happen.

PART II

OTHER WAYS TO AVOID CREATING A BAD SYSTEM

8

Smart Start

8

SMART START

This chapter concerns having a Systems Development Manual in your organization. Since every organization is unique, this chapter is not meant to be used as a manual but as a possible outline for one. The system development process is a structural approach to building information systems. SDP is simply a sequence of rational and logical thoughts and actions used regularly in making decisions. As such, it is an essential tool used to increase individual contributions in project planning and development and to ensure project success.

An automated project management system should parallel SDP. The automated project management system is used for planning, budgeting and monitoring every work effort—thereby eliminating the negative effects of developing systems. This reporting system is described in the next chapter.

The rest of the chapter will be a pictorial walk-through of the Systems Development Manual presented through eight figures.

Figure 8.1 Systems Development Manual – shows the different areas as a family that could use this manual.

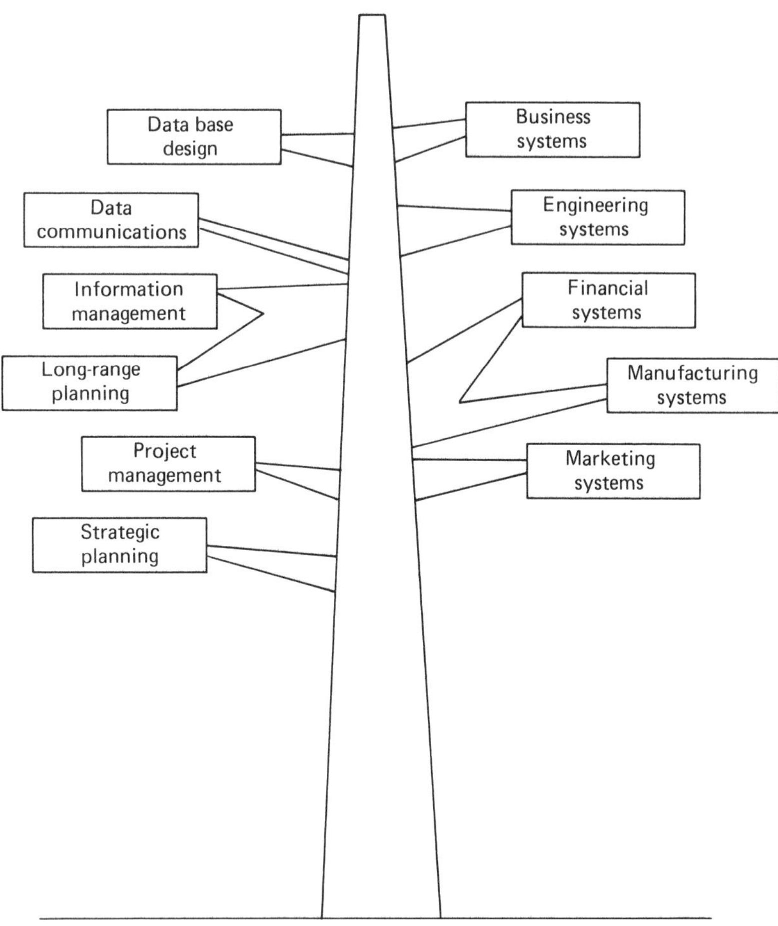

Figure 8.2 Risk/Cost Concept – shows the relationship of cumulative costs to the risks involved.

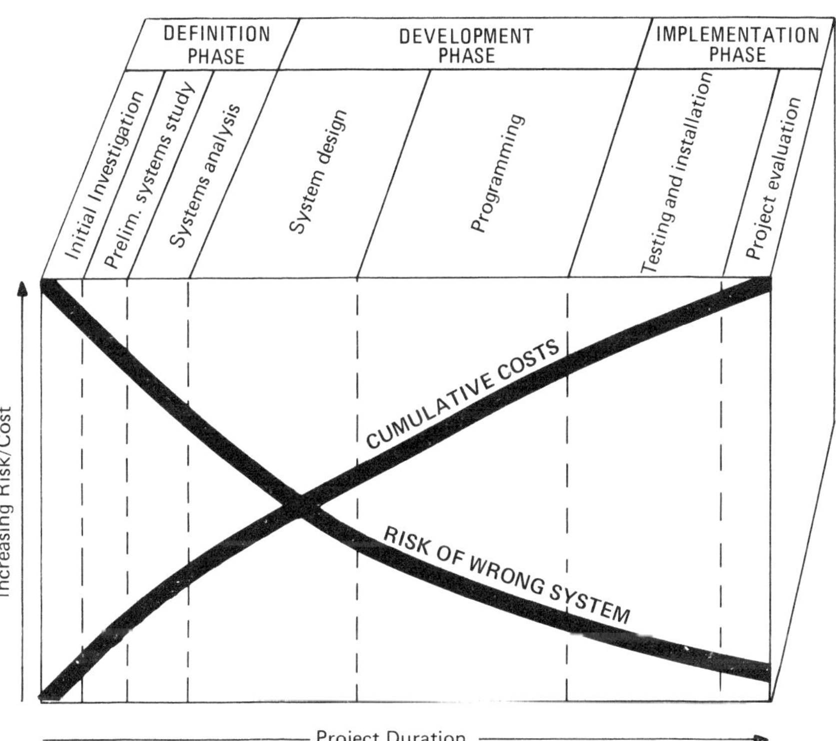

Figure 8.3 Gradual Management Steering Concept – shows the gradual commitment made by management for each step of the project.

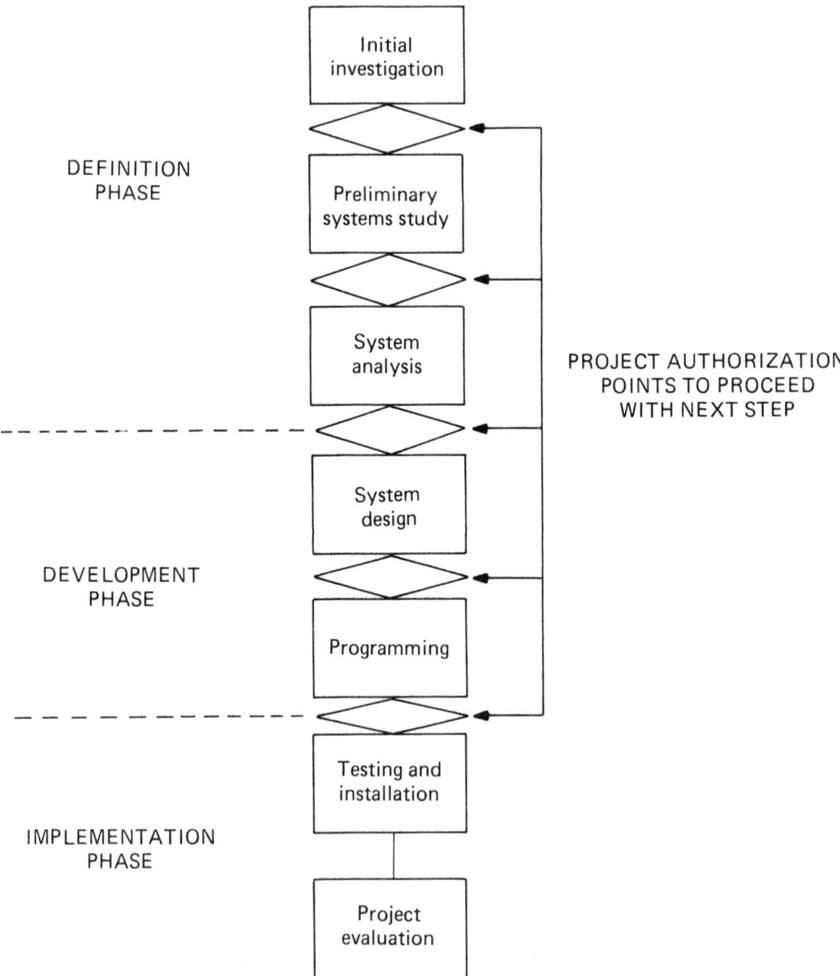

Figure 8.4. Systems Development Process (SDP) shows how to walk through each step of the project from the bottom to the top step.

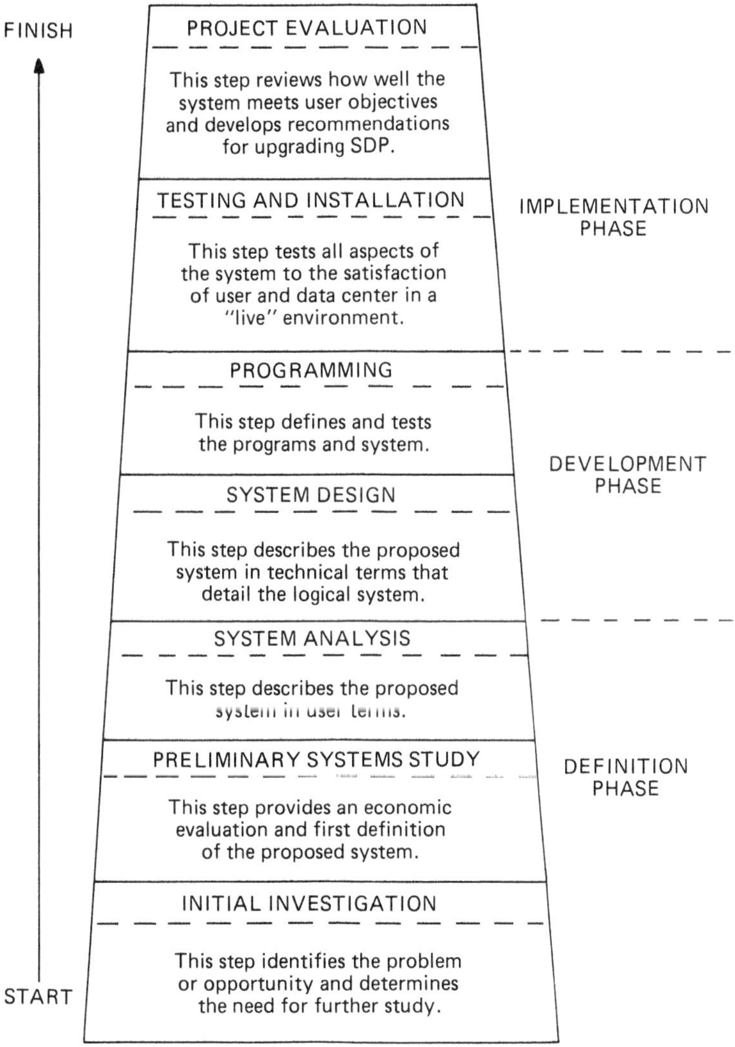

FINISH

PROJECT EVALUATION

This step reviews how well the system meets user objectives and develops recommendations for upgrading SDP.

TESTING AND INSTALLATION

This step tests all aspects of the system to the satisfaction of user and data center in a "live" environment.

IMPLEMENTATION PHASE

PROGRAMMING

This step defines and tests the programs and system.

SYSTEM DESIGN

This step describes the proposed system in technical terms that detail the logical system.

DEVELOPMENT PHASE

SYSTEM ANALYSIS

This step describes the proposed system in user terms.

PRELIMINARY SYSTEMS STUDY

This step provides an economic evaluation and first definition of the proposed system.

DEFINITION PHASE

INITIAL INVESTIGATION

This step identifies the problem or opportunity and determines the need for further study.

START

Figure 8.5 Logical Tasks for the Definition Phase—describes the initial investigation, preliminary systems study and system analysis steps of the SDP

Step	*Task Description*
Initial Investigation	1. Project authorization.
	2. Resolve objectives and opportunities.
	3. Prepare report stating recommendations, alternatives, anticipated benefits and/or results.
	4. Prepare a preliminary systems study plan (schedule and costs) for the next step of the project.
	5. Present report and plan to users and management to secure approval for the next step.
Preliminary Systems Study	6. Collect and review present system documentation and current costs.
	7. Identify input and output requirements.
	8. Develop high-level systems flow.
	9. Identify in or out of house application packages.
	10. Estimate cost and payback analysis.
	11. Prepare a system analysis plan for the next step of the project.
	12. Present plan to users and management to secure approval for the next step.
System Analysis	13. Perform data verification.
	14. Develop functional design, test and conversion strategy.

15. Review functional specifications with available application packages.

16. Prepare a system design plan for the next step of the project.

17. Present plan to users and management to secure approval for the next phase.

Figure 8.6 Logical Tasks for the Development Phase—describes the system design and programming steps of the SDP

Step	*Task Description*
System Design	18. Design and review data base.
	19. Identify hardware requirements.
	20. Identify test data requirements.
	21. Prepare detailed logical charts for each function.
	22. Finalize input and output requirements.
	23. Develop implementation/conversion plan and user procedures.
	24. Prepare a programming plan for the next step of the project.
	25. Present plan to users and management to secure approval for the next step.
Programming	26. Review systems specifications and determine program modules/job flow.
	27. Assign work to programmers.
	28. Review data base design and program specifications.

29. Prepare and walk through top-down diagram.

30. Prepare test data.

31. Code, debug and test programs.

32. Prepare program documentation, operations documentations and production JCL (job control language).

33. Test complete system.

34. Secure approval from management for the next phase.

Figure 8.7 Logical Tasks for the Implementation Phase—describes the testing/implementation and project evaluation steps of the SDP

Step	*Task Description*
Testing and Installation	35. Review conversion plan and systems test plan.
	36. Train users.
	37. Review user training manuals and computer operations documentation.
	38. Test complete system in a "live" environment.
	39. Verify results and conduct conversion.
	40. Obtain user acceptance.
Project Evaluation	41. Review how well the system meets user business objectives. This review should take place from one to six months after the system becomes operational.
	42. Develop recommendations for upgrading the systems development process.

Figure 8.8. General Roles/Responsibilities — describes the involvement of many categories of staff within the SDP

PHASE	*STEP*	*USER*		*SYSTEMS AND SERVICES*		*STEERING COMMITTEE*	*EDP AUDIT*	*SYSTEMS REVIEW COMMITTEE*
		MANAGEMENT	*STAFF*	*MANAGEMENT*	*STAFF*			
DEFINITION	Initial Investigation	AMR	P	R	CP	A	C	CR
	Preliminary Systems Study	RM	P	RM	P	A	C	CR
	System Analysis	RM	P	RM	P	A	C	CR
DEVELOPMENT	System Design	RM	P	RM	P	A	C	CR
	Programming			RM	P	A	C	CR
IMPLEMENTATION	Testing and Installation	RM	P	RM	P	A	C	CR
	Project Evaluation	C	P	C	P	R	C	MR

LEGEND:
A = Authorize
C = Consult or provide assistance
M = Manage and control
P = Perform
R = Review or evaluate

CONCLUSIONS

Having and using a Systems Development Manual in your organization will help you realize your objectives more effectively and sooner. The "successful" systems development process is the sum of the phased approach, Management Steering and the Automated Project Management System. The success of your system is the purpose and function of this manual and is a "smart start."

9

Meet JULIE and MICHAEL

9

MEET JULIE AND MICHAEL

FLOPS Flops (chapter 1) was about a project that did not meet its implementation dates, went over budget, was still being modified after implementation and was not a quality-built system. The story could have been rewritten as "FLOPS Tops" if the systems department had an automated project management system for planning, budgeting and monitoring every work effort. An automated project management system should have the following attributes:

On-line proprietary package for in-house installation. The package should be easily installed and used. Or uses the package on an outside time-sharing environment.

A system that is easy to learn, use and requires no prior programming experience. The system will use English-like commands.

Training and implementation support is provided.

Answers the crucial planning questions, what, who, how much and when.

Cost effective to use.

Reduces development and maintenance costs.

A good system for a development methodology which has a high degree of flexibility.

Handles a multiproject environment.

Summarizes project or projects.

Determines project trends.

Sets realistic work schedules.

"If what" planning and simulation capabilities.

"Done" and "to do" project steps and costs which reflect project history and future commitments.

Budget dollars showing monthly expenditures and schedules expenditures based on the specific budgeting areas and their functions.

Need to manipulate large amounts of data and projects.

Retains data on accessible output files.

Easy input and fast changes.

Editing and refining capabilities.

Includes pictorial charts.

There are only a handful of software packages available that meet the above requirements. The two packages that we shall look at an overview of are Julie and Michael. These packages are cost effective to us and are proven products.

JULIE

JULIE is a computerized project management system. Figure 9.1 shows the general overview of JULIE.

Figure 9.1. General overview of JULIE.

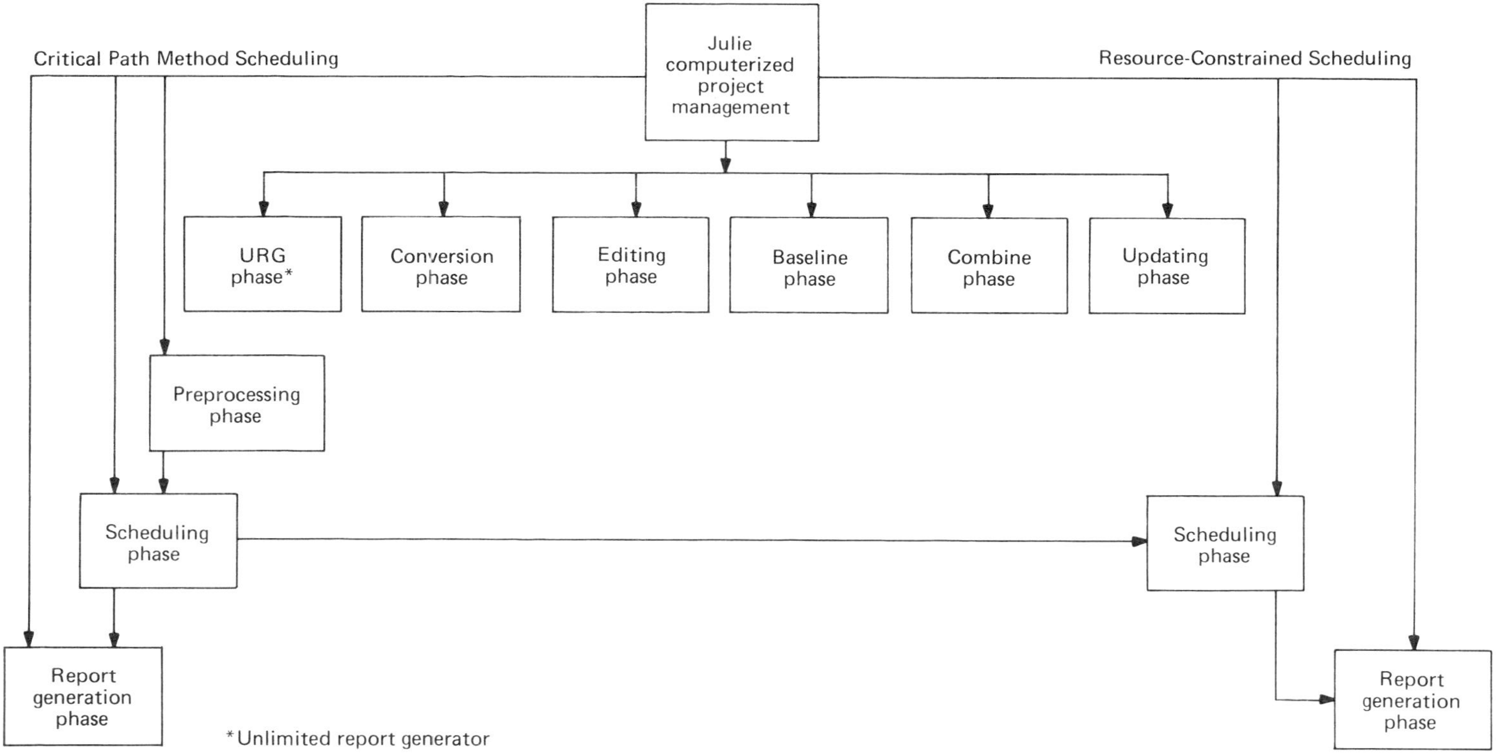

JULIE offers the following services:

Assign available people and machines to work on activities

Determine when the project can be finished and at what cost

Generate a variety of reports

Produce quickly and accurately several alternative schedules which include CPM (critical path method) scheduling

Schedule hundreds of activities according to their estimated duration and interdependencies

JULIE offers the following reports:
Bar charts

Combined resource

Cost reports

Distribution of resources reports

Logic reports

Network diagrams

Tabular reports

Utilization of resources reports

MICHAEL

MICHAEL is a project planning and control system. Figure 9.2 gives the general overview of MICHAEL.

Figure 9.2 General overview of MICHAEL.

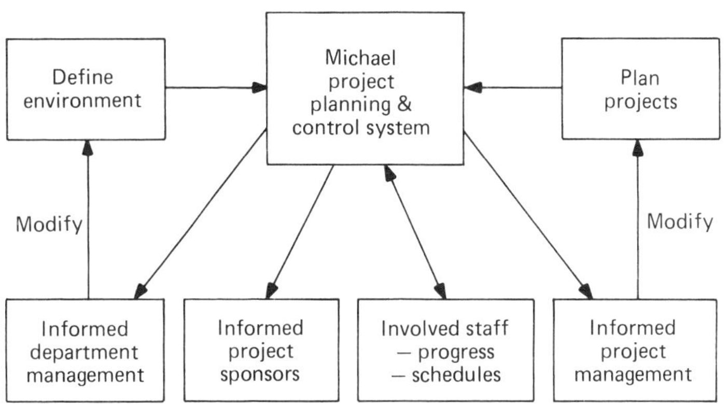

MICHAEL offers the following services:

Automated network management

Automated project generation

Both process and event networking

Critical path management

Daily automatic load leveling

Extensive data validation

Reporting for matrix organization

Substantial audit capability

Total resource accounting and commitments

Unlimited "what if" simulation

Variable resource load smoothing

MICHAEL offers the following reports:

Accounting
- billing
- budget

Administration
- dictionary
- transactions

History and commitment of resources
- category forecast by projects
- category history by projects
- project forecast by category
- project history by category

Performance analysis
- plan versus actual
- prediction analysis
- project evaluation
- task evaluation

Project planning and control
- bar charts
- project analysis
- project review by category
- network analysis
- trend analysis

Resource load and distribution
- resource allocation
- resource availability
- work schedule

CONCLUSIONS

Project management can be broken down into two cycles—planning and control. The planning cycle defines the project goals and produces a detailed schedule on how they will be achieved. Also, this cycle must be flexible enough to be revised many times. The control cycle is an ongoing process of monitoring and updating that does not end until the project is implemented.

An automated project management system should improve individual productivity and communications with the user community. The system will define the roles and responsibilities of the EDP and user participants. Also, the system will meet time, cost and objectives involved in the development of the project and at the same time maximize all resources.

Not using an automated project management system in today's large project environment is comparable to a man who started his own business from nothing. Within the first 10 years, his business grew from 1 to 20 million dollars in annual sales. Part of his success formula was to make all the decisions and not to delegate any authority. Within the next 5 years the annual sales increased to 100 million dollars. In the sixteenth year the owner had a nervous breakdown trying to run his company. The owner made the mistake of extending his proven control formula for a small company to a large company. Likewise, as the size of projects increases the ground rules must change to have a better control system. An automated project management system gives project managers a toolbox for a better control system. The most important aspect of using an automated management system is to eliminate the negative effects of developing projects.

10

INSIGHT

10

INSIGHT

Let us look at some possible alternatives to the traditional data processing life cycle for business applications. This chapter and the next will look at two different approaches.

There are dozens of financial modeling languages in use today, but relatively few are available as proprietary packages for in-house installation. The proprietary package that we shall look at here is called INSIGHT.

INSIGHT is a modeling and reporting system that can be used easily by accountants, business managers and other non-DP personnel to generate financial and management reports. The language that runs INSIGHT is a transparent computer language that is easy for a nonprogrammer to learn and use, and this tool is becoming increasingly flexible. The most common usage of INSIGHT is in a time-sharing environment which is the access to the computer from typewriter-like terminals (video display screens and/or keyboard printer terminals) over phone lines.

There are a number of advantages of using INSIGHT. First, the traditional business application can take six or more months to implement. The user works through the systems analyst; the systems analyst works through the programmer; the programmer works through computer operations. As one can see, this can be a very time-consuming and expensive business application for the user. Quite often, after the application is implemented, the user is not happy with the outputs and the above working cycle starts over for the modification. With INSIGHT, however, the user circumvents the user-systems analyst-programmer-computer operations cycle by setting up his or her own application. The major investment of the user using INSIGHT is his or her own time. It takes only minutes or a few hours to design and implement a business application. The main difference between an INSIGHT model and the traditional program is that the model combines the input process and the program has a separate input and process. See Figure 10.1.

Figure 10.1

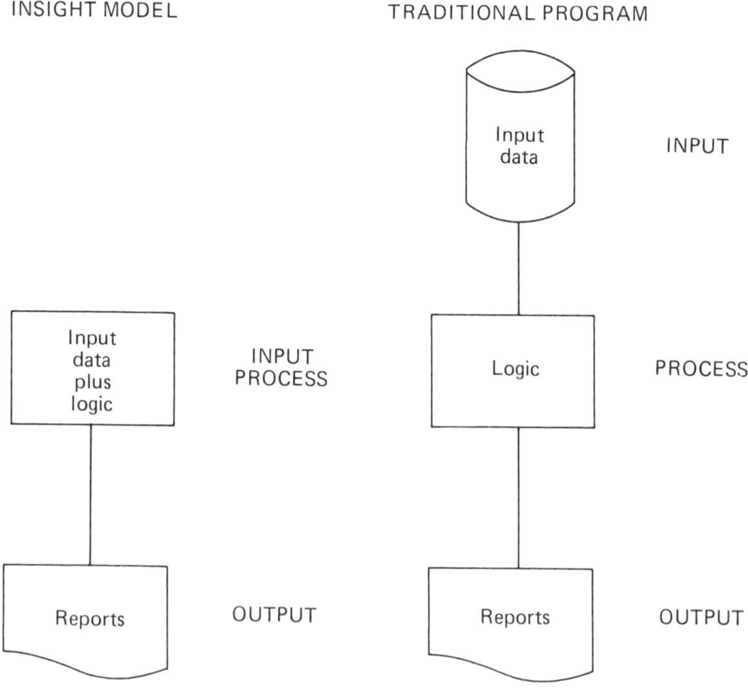

Second, INSIGHT is easy to learn and use by using only English instructions.

Third, you can have computer-printed reports in a matter of minutes that would take days or weeks to produce manually.

Fourth, your attention is focused on your reports and not on the computer.

Fifth, INSIGHT has a data base for storing and retrieving a number of business applications (models).

Sixth, INSIGHT has the inherent ability to change the reporting logic quickly (calculations, relationships and report formats).

Among the features of INSIGHT are the following:

1. Financial planning, forecasting, analysis, modeling and "what if" simulation to explore alternative solutions
2. Automated financial routines—AMORTIZE, ANNUITY, DEPRECIATE, DISCOUNT, PRESENT VALUE, RATE OF RETURN and SPREAD
3. Consolidation of models—CONSOLMERGE and CONSOLSELECT
4. Sophisticated reporting—REPORT GENERATOR/WRITER with variable page length and width
5. Plotting line or bar graphs

Before relating a specific business application, let us define a few INSIGHT terms:

Model—a user-defined business application which consists of a matrix in which the user defines the logic. The matrix is made up of lines and columns and their interrelationships. The model contains column headings, line numbers with names and definitions of reports to be printed. Each model can have up to ten different reports.

Disk files—a file that permanently stores repetitive commands instead of wasting time entering each individual command. Also, there will be less potential to make errors. The command to execute the disk file is INPUT DISK NAME-OF-FILE.

Report generator—the standard report format found within a model may need some fine tuning or window dressing. Using this feature, the user creates a file containing the report editing and formatting instructions. The command to execute the report generator after the model is called into memory is:

WRITER
NAME-OF-FILE

CASE IN POINT

The consolidation of branch profit and loss statements for the Smith Transport Company will be used as an example. The systems overview is broken down into four areas—input models, disk files, output models and output reports. See Figure 10.2.

Figure 10.2 Branch profit and loss statements – systems overview

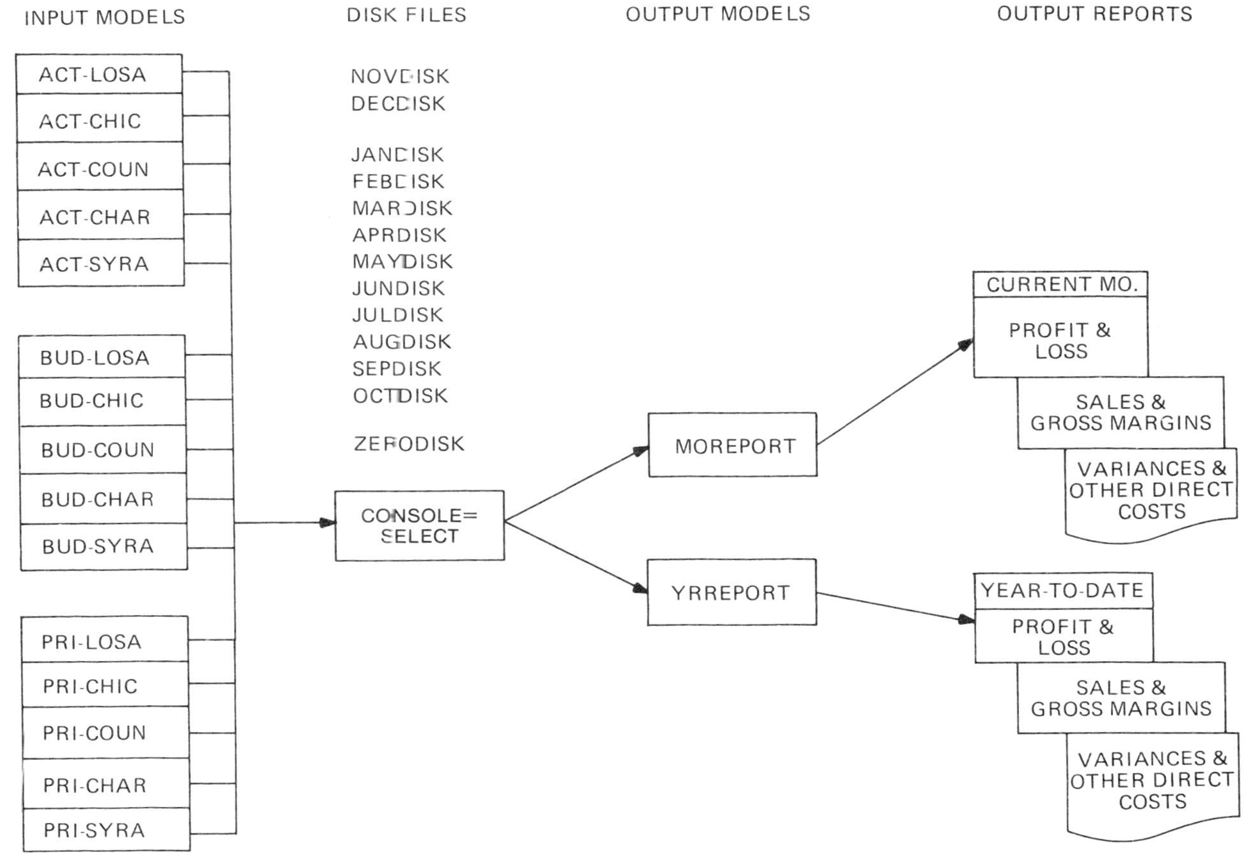

Input Models

The five branches owned and operated by the Smith Transport Company are located in Los Angeles, Chicago, Council Bluffs, Charlotte and Syracuse. The following are the names of the input models.

Actual	*Budget*	*Prior year*
ACT-LOSA	BUD-LOSA	PRI-LOSA
ACT-CHIC	BUD-CHIC	PRI-CHIC
ACT-COUN	BUD-COUN	PRI-COUN
ACT-CHAR	BUD-CHAR	PRI-CHAR
ACT-SYRA	BUD-SYRA	PRI-SYRA

The actual models are updated on a monthly basis and the budget/prior year models are updated on a yearly basis.

See Figure 10.3 for a listing (DISPLAY SOURCE) of ACT-LOSA. With the exception of model name, report title and identification (ID), all fifteen input models have the same lines and logic. All fifteen input models were created from one master input model. Using a master input model saves a lot of time versus setting up fifteen individual input models.

Below is the standard report format always found in a DISPLAY SOURCE:

1. PE or PERIODS command is used to set the number of columns to be used within the model matrix. This should be the first command to be used in building a model.
2. PAGE command applies to the printing of reports. The PAGE parameters are used to define the number of lines per page, the maximum number of print positions across the page, the number of columns per report page and how many print characters per column and line name.

3. HE or HEADINGS command is used to insert titles over all columns found within a model.
4. RE or REPORT command defines the line number range, column number range and the report titles.
5. DATE command is the input date or time frame that the model is being used. This information appears on the upper left-hand corner of each page.
6. ID or IDENTIFICATION command is an optional field that can be used for any purpose. This information appears on the upper right-hand corner of each page. A good habit to get into is to have the model name be the ID.
7. L or LINE definitions are made up of three types of lines: title lines only, input data lines (have zero values in Figure 10.3) and relationship (using equal signs) lines.
8. COL or COLUMN definitions are made up of column relationships.

Disk Files

There is a disk file for each month of the year. The first month of the fiscal year for the Smith Transport Company is November. The following are the names of the monthly disk files:

NOVDISK	FEBDISK	MAYDISK	AUGDISK
DECDISK	MARDISK	JUNDISK	SEPDISK
JANDISK	APRDISK	JULDISK	OCTDISK

See Figure 10.4 for a listing of JANDISK. With the exception of disk file name and column number (column 3 represents January and column 15 represents January year-to-date) commands, all twelve disk files have the same lines and logic. All twelve monthly disk files were created from one master disk file versus seting up twelve individual disk files.

The first-half logic of the JANDISK disk file is to select column 3 (January) from the fifteen input models and save the data in an output model called MOREPORT. After this is completed, the monthly output reports are printed.

The second-half logic is a mirror image of the first-half logic. The exception is that column 15 (January year-to-date) is being selected and saved in an output model called YRREPORT. After this is completed, the year-to-date output reports are printed.

There is a thirteenth disk file called ZERODISK. This file is executed after the last yearly reports are printed in October and before the November updating of the five actual input models. This disk file moves the data from the five actual input models into its corresponding prior year input models. The next step is to zero-out the five actual input models and the five budget input models. The last step is to enter manually the yearly budget data into the five budget input models.

Output Models

In Figure 10.5, part (b) is a continuation of part (a), and part (c) is a continuation of part (b). The listing is a DISPLAY SOURCE of the output model called YRREPORT. With the exception of model name, report titles and identification (ID), both output models have the same lines and logic. Also, both output models were created from one master output model.

The standard report format is the same as the input models. YRREPORT and MOREPORT output models both have the same three reports defined within each model.

Output Reports

The following are the six output report titles:

"Profit and Loss"—monthly and year-to-date
"Sales and Gross Margins"—monthly and year-to-date
"Variances and Other Direct Costs"—monthly and year-to-date.

See Figure 10.6 for a consolidated output report.

Figure 10.7 is a listing of a report generator that is used to edit the six output reports. The editing is used to change the standard report format found within the output models. This listing defines the heading titles for each column, column widths, commas to separate thousands and zero-suppression (leaving a blank field) of values from printing.

BENEFITS

Before the INSIGHT system was implemented, it took users forty manual hours per month to produce the Branch Profit and Loss Statements. Now it takes four hours a month using INSIGHT for a 90% reduction in time. Also, the computer reports look more professional versus handwritten reports and were done right the first time.

The two main benefits of using INSIGHT are the savings in manual time and the fact that it is a great management tool. Additionally, INSIGHT systems are flexible enough to grow with.

INSIGHT language has been in use for many years and has continuously been maintained and improved to meet user demands. Approximately each year, INSIGHT comes out with a new version of their software package. The new version is made up of new commands, enhancements to pre-existing commands and better processing efficiency, making INSIGHT a good proprietary software package that improves every year.

Figure 10.3

```
DISPLAY SOURCE
FILENAME:   ACT-LOSA

PE - 23, MAX NO. LINES - 458 CURR. NO. LINES - 39 CURR. NO. RELS - 1
PAGE LENGTH 0 WIDTH 130 COLS 13 CHARS 8 NAMES 16 MESS
HE  T  COL 1 '        NOV' '        DEC' '        JAN' '        FEB' '        MAR' '        APR' >
'     MAY' '     JUN' '     JUL' '     AUG'
HE  T  COL 11 '     SEP' '     OCT' '     TOTAL YR'     'DEC YTD '     'JAN YTD'     'FEB YTD' >
'MAR YTD'     'APR YTD'     'MAY YTD'     'JUN YTD'
HE  T  COL 21     'JUL YTD'     'AUG YTD'     'SEP YTD'
RE  1   L   1   T   999.9   COL 1   T   13   'ACTUAL INPUT'     'FOR LOS ANGELES'
DATE '4-1-78
ID 'ACT-LOSA
L   29.8   'VARIABLE INPUT'
L   29.9
L   30   'SHRINKAGE' 0 0 0 0 0 0 0 0 0 0 0 0 0 0 0 0 0 0 0 0 0 0 0
L   50   'OBSOLESCENCE' 0 0 0 0 0 0 0 0 0 0 0 0 0 0 0 0 0 0 0 0 0 0 0
L   70   'CHANGE IN STDS' 0 0 0 0 0 0 0 0 0 0 0 0 0 0 0 0 0 0 0 0 0 0 0
L   150 'PURCHASE VARIABL' 0 0 0 0 0 0 0 0 0 0 0 0 0 0 0 0 0 0 0 0 0 0 0
L   170 'RATE CHANGES' 0 0 0 0 0 0 0 0 0 0 0 0 0 0 0 0 0 0 0 0 0 0 0
L   190 'COST OVER STD RE' 0 0 0 0 0 0 0 0 0 0 0 0 0 0 0 0 0 0 0 0 0 0 0
L   269.7
L   269.8 'SALES INPUT'
L   269.9
L   270 'UNITS-TRAILERS' 0 0 0 0 0 0 0 0 0 0 0 0 0 0 0 0 0 0 0 0 0 0 0
L   275 'TRUCKS' 0 0 0 0 0 0 0 0 0 0 0 0 0 0 0 0 0 0 0 0 0 0 0
L   280 'USED' 0 0 0 0 0 0 0 0 0 0 0 0 0 0 0 0 0 0 0 0 0 0 0
L   300 'SRO - LABOR' 0 0 0 0 0 0 0 0 0 0 0 0 0 0 0 0 0 0 0 0 0 0 0
L   305 'INSTALLATION' 0 0 0 0 0 0 0 0 0 0 0 0 0 0 0 0 0 0 0 0 0 0 0
L   310 'PARTS' 0 0 0 0 0 0 0 0 0 0 0 0 0 0 0 0 0 0 0 0 0 0 0
```

```
L   330 'PARTS - DEALERS' 0 0 0 0 0 0 0 0 0 0 0 0 0 0 0 0 0 0 0 0 0 0 0
L   335 'BRANCH CUST' 0 0 0 0 0 0 0 0 0 0 0 0 0 0 0 0 0 0 0 0 0 0 0
L   340 'BRANCH DEALERS' 0 0 0 0 0 0 0 0 0 0 0 0 0 0 0 0 0 0 0 0 0 0 0
L   394 'GROSS MARGINS'
L   395 'UNITS-TRAILERS' 0 0 0 0 0 0 0 0 0 0 0 0 0 0 0 0 0 0 0 0 0 0 0
L   405 'TRUCKS' 0 0 0 0 0 0 0 0 0 0 0 0 0 0 0 0 0 0 0 0 0 0 0
L   415 'USED' 0 0 0 0 0 0 0 0 0 0 0 0 0 0 0 0 0 0 0 0 0 0 0
L   450 'SRO - LABOR' 0 0 0 0 0 0 0 0 0 0 0 0 0 0 0 0 0 0 0 0 0 0 0
L   460 'INSTALLATIONS' 0 0 0 0 0 0 0 0 0 0 0 0 0 0 0 0 0 0 0 0 0 0 0
L   470 'PARTS' 0 0 0 0 0 0 0 0 0 0 0 0 0 0 0 0 0 0 0 0 0 0 0
L   505 'PARTS-DEALERS' 0 0 0 0 0 0 0 0 0 0 0 0 0 0 0 0 0 0 0 0 0 0 0
L   515 'BRANCH CUST' 0 0 0 0 0 0 0 0 0 0 0 0 0 0 0 0 0 0 0 0 0 0 0
L   525 'BRANCH DEALERS' 0 0 0 0 0 0 0 0 0 0 0 0 0 0 0 0 0 0 0 0 0 0 0
L   749.7
L   749.8 'P & L INPUT'
L   749.9
L   750 'BR EXP - GROSS' 0 0 0 0 0 0 0 0 0 0 0 0 0 0 0 0 0 0 0 0 0 0 0
L   770 'SRO BURDEN ABSOR' 0 0 0 0 0 0 0 0 0 0 0 0 0 0 0 0 0 0 0 0 0 0 0
L   790 'GEN & ADMIN' 0 0 0 0 0 0 0 0 0 0 0 0 0 0 0 0 0 0 0 0 0 0 0
L   850 'OTHER INCOME NET' 0 0 0 0 0 0 0 0 0 0 0 0 0 0 0 0 0 0 0 0 0 0 0
L   860 '10% HAND CHG WAR' 0 0 0 0 0 0 0 0 0 0 0 0 0 0 0 0 0 0 0 0 0 0 0
L   999 'GRAND TOTAL = L   30   T   860
COL   13 = COL   1   T   12
COL   14 = COL   1   T   2
COL   15 = COL   1   T   3
COL   16 = COL   1   T   4
COL   17 = COL   1   T   5
COL   18 = COL   1   T   6
COL   19 = COL   1   T   7
COL   20 = COL   1   T   8
COL   21 = COL   1   T   9
COL   22 = COL   1   T   10
COL   23 = COL   1   T   11
```

Figure 10.4

```
JANDISK   LIST
  100 CONSOLSELECT FORMAT MOREPORT
  110 ACT-LOSA
  120 3   1
  130 NOMORE
  140 BUD-LOSA
  150 3   2
  160 NOMORE
  170 PRI-LOSA
  180 3   3
  190 NOMORE
  200 ACT-CHIC
  210 3   4
  220 NOMORE
  230 BUD-CHIC
  240 3   5
  250 NOMORE
  260 PRI-CHIC
  270 3   6
  280 NOMORE
  290 ACT-COUN
  300 3   7
  310 NOMORE
  320 BUD-COUN
  330 3   8
  340 NOMORE
```

```
570 CONSOLSELECT FORMAT YRREPORT
580 ACT-LOSA
590 15   1
600 NO MORE
610 BUD-LOSA
620 15   2
630 NOMORE
640 PRI-LOSA
650 15   3
660 NOMORE
670 ACT-CHIC
680 15   4
690 NOMORE
700 BUD-CHIC
710 15   5
720 NOMORE
730 PRI-CHIC
740 15   6
750 NOMORE
760 ACT-COUN
770 15   7
780 NOMORE
790 BUD-COUN
800 15   8
810 NOMORE
```

```
350 PRI-COUN
360 3   9
370 NOMORE
380 ACT-CHAR
390 3   10
400 NOMORE
410 BUD-CHAR
420 3   11
430 NOMORE
440 PRI-CHAR
450 3   12
460 NOMORE
461 ACT-SYRA
462 3   13
463 NO MORE
464 BUD-SYRA
465 3   14
466 NO MORE
467 PRI-SYRA
468 3   15
469 NO MORE
470 NOMORE
500 REPLACE >
510 MOREPORT
520 WRITER CALCULATE 2 TIMES
530 P&LRPT2   REPORT 1
540 P&LRPT2   REPORT 2
550 P&LRPT2   REPORT 3
560 NOMORE
```

```
820 PRI-COUN
830 15   9
840 NOMORE
850 ACT-CHAR
860 15   10
870 NOMORE
880 BUD-CHAR
890 15   11
900 NOMORE
910 PRI-CHAR
920 15   12
930 NOMORE
931 ACT-SYRA
932 15   13
933 NO MORE
934 BUD-SYRA
935 15   14
936 NO MORE
937 PRI-SYRA
938 15   15
939 NO MORE
940 NOMORE
970 REPLACE >
980 YRREPORT
990 WRITER CALCULATE 2 TIMES
1000 P&LRPT2   REPORT 1
1010 P&LRPT2   REPORT 2
1020 P&LRPT2   REPORT 3
1030 NOMORE
```

Figure 10.5a

```
DISPLAY SOURCE
FILENAME: YRREPORT

PE - 18, MAX NO. LINES - 578 CURR. NO. LINES - 121 CURR. NO. RELS - 34
PAGE LENGTH   0 WIDTH   130 COLS   13 CHARS   8 NAMES   16 MESS
HE T COL 1 '----L'     'OS ANGEL'     'ES----'     '------'     '-CHICAGO'     '------' >
'---COU'     'NCIL BLU'     'FFS---'     '------'
HE T COL 11 'CHARLOTT'     'E----'     '----CO'     'NSOLIDAT'     'ED----'
HE M COL 1  ' ACTUAL '  ' BUDGET '  ' PRIOR YR '  ' ACTUAL '  ' BUDGET '  'PRIOR YR' >
' ACTUAL '     ' BUDGET '     'PRIOR YR'     ' ACTUAL '
HE M COL 11   ' BUDGET '     ' PRIOR YR '     ' ACTUAL '     'BUDGET'     'PRIOR YR'
RE 1 L 1 T 250 COL 1 T 15 '        SMITH TRANSPORT COMPANY ' >
'VARIANCES AND OTHER DIRECT COSTS'
RE 2 L 250 T 600 COL 1 T 15 '       SMITH TRANSPORT COMPANY' >
' SALES AND GROSS MARGINS '
RE 3 L 601 T 999.9 COL 1 T 15 '        SMITH TRANSPORT COMPANY' >
'  PROFIT AND LOSS'
DATE 'FOR JUN 1979 YTD'
ID 'YRREPORT         '
L  1   'VARIANCE REPORT' SUP
L  10  'OTHER DIRECT:'
L  20
L  30  '  SHRINKAGE' 0 0 0 0 0 0 0 0 0 0 0 0 0 0 0 0 0 0
L  40
L  50  '  OBSOLESCENCE' 0 0 0 0 0 0 0 0 0 0 0 0 0 0 0 0 0 0
L  60
L  70  '  CHANGE IN STDS' 0 0 0 0 0 0 0 0 0 0 0 0 0 0 0 0 0 0
L  80
L  90  '-'
L  100 'TOT OTHER DIRECT' = L 30 + L 50 + L 70
L  110
```

```
L  120
L  130 'VARIANCES:'
L  140
L  150 '   PURCHASE' 0 0 0 0 0 0 0 0 0 0 0 0 0 0 0 0 0 0
L  160
L  170 '   RATE CHANGES' 0 0 0 0 0 0 0 0 0 0 0 0 0 0 0 0 0 0
L  180
L  190 '   OVER STD REBLD' 0 0 0 0 0 0 0 0 0 0 0 0 0 0 0 0 0 0
L  200
L  210 '-'
L  220 ' TOTAL VARIANCES' = L 150 + L 170 + L 190
L  230 '='
L  251 'SALES REPORT' SUP
L  260 'GROSS SALES:'
L  265
L  270 'UNITS-TRAILERS' 0 0 0 0 0 0 0 0 0 0 0 0 0 0 0 0 0 0
L  275 '          -TRUCKS' 0 0 0 0 0 0 0 0 0 0 0 0 0 0 0 0 0 0
L  280 '          -USED' 0 0 0 0 0 0 0 0 0 0 0 0 0 0 0 0 0 0
L  285 '-'
L  290 'TOTAL UNITS' = L 270 + L 275 + L 280
L  295
L  300 'SRO-LABOR' 0 0 0 0 0 0 0 0 0 0 0 0 0 0 0 0 0 0
L  305 '      -INSTALLATION' 0 0 0 0 0 0 0 0 0 0 0 0 0 0 0 0 0 0
L  310 '      -PARTS' 0 0 0 0 0 0 0 0 0 0 0 0 0 0 0 0 0 0
L  315 '-'
L  320 'TOTAL SRO' = L 300 + L 305 + L 310
L  325
L  330 'PARTS-DEALERS' 0 0 0 0 0 0 0 0 0 0 0 0 0 0 0 0 0 0
L  335 '          -BR CUST' 0 0 0 0 0 0 0 0 0 0 0 0 0 0 0 0 0 0
L  340 '          -BR DEALERS' 0 0 0 0 0 0 0 0 0 0 0 0 0 0 0 0 0 0
L  345 '-'
L  350 'TOTAL PARTS' = L 330 + L 335 + L 340
L  355 '-'
L  360 'TOTAL SALES' = L 290 + L 320 + L 350
```

Figure 10.5b

```
L  370
L  375
L  380
L  385 'GROSS MARGINS:'
L  390
L  395 'UNITS-TRAILERS' 0 0 0 0 0 0 0 0 0 0 0 0 0 0 0 0 0 0
L  400 '      -% OF SALES'   DEC 1 =  100 * L 395 / L 270
L  405 '          -TRUCKS' 0 0 0 0 0 0 0 0 0 0 0 0 0 0 0 0 0 0
L  410 '      -% OF SALES'  DEC 1 =  100 * L 405 / L 275
L  415 '          -USED' 0 0 0 0 0 0 0 0 0 0 0 0 0 0 0 0 0 0
L  420 '      -% OF SALES'  DEC 1 =  100 * L 415 / L 280
L  425 '-'
L  430 'TOTAL UNITS' = L 395 + L 405 + L 415
L  435 '      -% OF SALES'   DEC 1 = 100 * L 430 / L 290
L  440 '-'
L  445
L  450 'SRO-LABOR' 0 0 0 0 0 0 0 0 0 0 0 0 0 0 0 0 0 0
L  455 '      -% OF SALES'  DEC 1 =  100 * L 450 / L 300
L  460 '      -INSTALLATION' 0 0 0 0 0 0 0 0 0 0 0 0 0 0 0 0 0 0
L  465 '      -% OF SALES'  DEC 1 =  100 * L 460 / L 305
L  470 '      -PARTS' 0 0 0 0 0 0 0 0 0 0 0 0 0 0 0 0 0 0
L  475 '      -% OF SALES'  DEC 1 =  100 * L 470 / L 310
L  480 '-'
L  485 'TOTAL SRO' = L 450 + L 460 + L 470
L  490 '      -% OF SALES'  DEC 1 =  100 * L 485 / L 320
L  495 '-'
L  500
L  505 'PARTS-DEALERS' 0 0 0 0 0 0 0 0 0 0 0 0 0 0 0 0 0 0
L  510 '      -% OF SALES'  DEC 1 =  100 * L 505 / L 330
L  515 '          -BR CUST' 0 0 0 0 0 0 0 0 0 0 0 0 0 0 0 0 0 0
L  520 '      -% OF SALES'  DEC 1 =  100 * L 515 / L 335
L  525 '          -BR DEALERS' 0 0 0 0 0 0 0 0 0 0 0 0 0 0 0 0 0 0
L  530 '      -% OF SALES'  DEC 1 =  100 * L 525 / L 340
L  535 '-'
L  540 'TOTAL PARTS' = L 505 + L 515 + L 525
L  545 '      -% OF SALES'  DEC 1 =  100 * L 540 / L 350
L  550 '-'
```

```
L  555
L  560 '-'
L  565 'TOTAL GROSS MARG' = L 430 + L 485 + L 540
L  570 '     -% OF SALES' DEC 1 = 100 * L 565 / L 360
L  575 '-'
L  601 'P & L REPORT'   SUP
L  610 'GROSS SALES' = L 360
L  620
L  630 'GROSS MARGIN-STD' = L 565
L  640 '      % OF SALES' DEC 1 = 100 * L 630 / L 610
L  650 '-'
L  660
L  670 'OTHER DIRECT' = L 100
L  680
L  690 'VARIANCES' = L 220
L  700 '-'
L  710 '      'GROSS MARGIN-ACT' = L 630 + L 670 + L 690
L  720 '      % OF SALES'   DEC 1 = 100 * L 710 / L 610
L  730 '-'
L  740
L  750 'BRANCH EXP-GROSS' 0 0 0 0 0 0 0 0 0 0 0 0 0 0 0 0 0 0
L  760
L  770 'SRO BURDEN ABSOR' 0 0 0 0 0 0 0 0 0 0 0 0 0 0 0 0 0 0
L  780
L  790 'GEN & ADMIN-2.3%' 0 0 0 0 0 0 0 0 0 0 0 0 0 0 0 0 0 0
L  800 '-'
L  810 'TOTAL EXP-NET' = L 750 + L 770 + L 790
L  820 '      % OF SALES' DEC 1 = 100 * L 810 / L 610
L  840
L  850 'OTHER INCOME-NET' 0 0 0 0 0 0 0 0 0 0 0 0 0 0 0 0 0 0
L  855
L  860 '15% HAND CHG-WAR' 0 0 0 0 0 0 0 0 0 0 0 0 0 0 0 0 0 0
L  865 '-'
L  870 'P&L BEFORE TAXES' = L 710 - L 810 + L 850 + L 860
L  880 '      % OF SALES' DEC 1 = 100 * L 870 / L 610
L  890 '-'
```

Figure 10.5c

```
COL 16 L 1   T 395 = COL 1 + COL 4 + COL 7 + COL 10 + COL 13
COL 16 L 405 T 405 = COL 1 + COL 4 + COL 7 + COL 10 + COL 13
COL 16 L 415 T 415 = COL 1 + COL 4 + COL 7 + COL 10 + COL 13
COL 16 L 425 T 430 = COL 1 + COL 4 + COL 7 + COL 10 + COL 13
COL 16 L 440 T 450 = COL 1 + COL 4 + COL 7 + COL 10 + COL 13
COL 16 L 460 T 460 = COL 1 + COL 4 + COL 7 + COL 10 + COL 13
COL 16 L 470 T 470 = COL 1 + COL 4 + COL 7 + COL 10 + COL 13
COL 16 L 480 T 485 = COL 1 + COL 4 + COL 7 + COL 10 + COL 13
COL 16 L 495 T 505 = COL 1 + COL 4 + COL 7 + COL 10 + COL 13
COL 16 L 515 T 515 = COL 1 + COL 4 + COL 7 + COL 10 + COL 13
COL 16 L 525 T 525 = COL 1 + COL 4 + COL 7 + COL 10 + COL 13
COL 16 L 535 T 540 = COL 1 + COL 4 + COL 7 + COL 10 + COL 13
COL 16 L 550 T 565 = COL 1 + COL 4 + COL 7 + COL 10 + COL 13
COL 16 L 575 T 630 = COL 1 + COL 4 + COL 7 + COL 10 + COL 13
COL 16 L 650 T 710 = COL 1 + COL 4 + COL 7 + COL 10 + COL 13
COL 16 L 730 T 810 = COL 1 + COL 4 + COL 7 + COL 10 + COL 13
COL 16 L 830 T 870 = COL 1 + COL 4 + COL 7 + COL 10 + COL 13
COL 17 L 1   T 395 = COL 2 + COL 5 + COL 8 + COL 11 + COL 14
COL 17 L 405 T 405 = COL 2 + COL 5 + COL 8 + COL 11 + COL 14
COL 17 L 415 T 415 = COL 2 + COL 5 + COL 8 + COL 11 + COL 14
COL 17 L 425 T 430 = COL 2 + COL 5 + COL 8 + COL 11 + COL 14
COL 17 L 440 T 450 = COL 2 + COL 5 + COL 8 + COL 11 + COL 14
COL 17 L 460 T 460 = COL 2 + COL 5 + COL 8 + COL 11 + COL 14
COL 17 L 470 T 470 = COL 2 + COL 5 + COL 8 + COL 11 + COL 14
```

```
COL 17 L 480 T 485 = COL 2 + COL 5 + COL 8 + COL 11 + COL 14
COL 17 L 495 T 505 = COL 2 + COL 5 + COL 8 + COL 11 + COL 14
COL 17 L 515 T 515 = COL 2 + COL 5 + COL 8 + COL 11 + COL 14
COL 17 L 525 T 525 = COL 2 + COL 5 + COL 8 + COL 11 + COL 14
COL 17 L 535 T 540 = COL 2 + COL 5 + COL 8 + COL 11 + COL 14
COL 17 L 550 T 565 = COL 2 + COL 5 + COL 8 + COL 11 + COL 14
COL 17 L 575 T 630 = COL 2 + COL 5 + COL 8 + COL 11 + COL 14
COL 17 L 650 T 710 = COL 2 + COL 5 + COL 8 + COL 11 + COL 14
COL 17 L 730 T 810 = COL 2 + COL 5 + COL 8 + COL 11 + COL 14
COL 17 L 830 T 870 = COL 2 + COL 5 + COL 8 + COL 11 + COL 14
COL 18 L 1   T 395 = COL 3 + COL 6 + COL 9 + COL 12 + COL 15
COL 18 L 405 T 405 = COL 3 + COL 6 + COL 9 + COL 12 + COL 15
COL 18 L 415 T 415 = COL 3 + COL 6 + COL 9 + COL 12 + COL 15
COL 18 L 425 T 430 = COL 3 + COL 6 + COL 9 + COL 12 + COL 15
COL 18 L 440 T 450 = COL 3 + COL 6 + COL 9 + COL 12 + COL 15
COL 18 L 460 T 460 = COL 3 + COL 6 + COL 9 + COL 12 + COL 15
COL 18 L 470 T 470 = COL 3 + COL 6 + COL 9 + COL 12 + COL 15
COL 18 L 480 T 485 = COL 3 + COL 6 + COL 9 + COL 12 + COL 15
COL 18 L 495 T 505 = COL 3 + COL 6 + COL 9 + COL 12 + COL 15
COL 18 L 515 T 515 = COL 3 + COL 6 + COL 9 + COL 12 + COL 15
COL 18 L 525 T 525 = COL 3 + COL 6 + COL 9 + COL 12 + COL 15
COL 18 L 535 T 540 = COL 3 + COL 6 + COL 9 + COL 12 + COL 15
COL 18 L 550 T 565 = COL 3 + COL 6 + COL 9 + COL 12 + COL 15
COL 18 L 575 T 630 = COL 3 + COL 6 + COL 9 + COL 12 + COL 15
COL 18 L 650 T 710 = COL 3 + COL 6 + COL 9 + COL 12 + COL 15
COL 18 L 730 T 810 = COL 3 + COL 6 + COL 9 + COL 12 + COL 15
COL 18 L 830 T 870 = COL 3 + COL 6 + COL 9 + COL 12 + COL 15
```

Figure 10.6

SMITH TRANSPORT COMPANY
VARIANCES AND OTHER DIRECT COSTS

FOR JUN 1979 YTD | YRREPORT

LINE NO		LOS ANGELES			CHICAGO			COUNCIL BLUFFS		
		ACTUAL	BUDGET	PRIOR YR	ACTUAL	BUDGET	PRIOR YR	ACTUAL	BUDGET	PRIOR YR
10.0	OTHER DIRECT:									
20.0										
30.0	SHRINKAGE	-3,235	-3,235	-3,210	-1,941	-1,941	-1,270	-1,294	-1,294	-1,900
40.0										
50.0	OBSOLESCENCE	-3,330	-3,235	-3,250	-1,998	-1,941	-1,300	-1,998	-1,941	-2,966
60.0										
70.0	CHANGE IN STDS	9,220	5,000	6,464	4,393	4,000	3,665	8,499	4,500	3,831
80.0										
90.0										
100.0	TOT OTHER DIRECT	2,655	-1,470	4	454	118	1,095	5,207	1,265	-1,035
110.0										
120.0										
130.0	VARIANCES:									
140.0										
150.0	PURCHASE	-956	-328	312	-982	200	-796	-244	1,664	1,306
160.0										
170.0	RATE CHANGES									
180.0										
190.0	OVER STD REBLD	-1,376	-3,664	-4,689	-1,002	-1,328	-1,272	-735	-328	-305
200.0										
210.0										
220.0	TOTAL VARIANCES	-2,332	-3,992	-4,377	-1,984	-1,128	-2,068	-979	1,336	1,001
230.0										

SMITH TRANSPORT COMPANY
VARIANCES AND OTHER DIRECT COSTS

FOR JUN 1979 YTD YRREPORT

LINE NO		CHARLOTTE ACTUAL	CHARLOTTE BUDGET	CHARLOTTE PRIOR YR	SYRACUSE ACTUAL	SYRACUSE BUDGET	SYRACUSE PRIOR YR	CONSOLIDATED ACTUAL	CONSOLIDATED BUDGET	CONSOLIDATED PRIOR YR
10.0	OTHER DIRECT:									
20.0										
30.0	SHRINKAGE	-1,941	-1,941	-1,270	-1,500	-1,500		-9,911	-9,911	-7,650
40.0										
50.0	OBSOLESCENCE	-666	-647	-1,300	-500	-500		-8,492	-8,264	-8,816
60.0										
70.0	CHANGE IN STDS	4,349	2,000	1.198				26,461	15,500	15,158
80.0										
90.0										
100.0	TOT OTHER DIRECT	1,742	-588	-1,372	-2,000	-2,000		8,058	-2,675	-1,308
110.0										
120.0										
130.0	VARIANCES:									
140.0										
150.0	PURCHASE	23	-664					-2,159	872	822
160.0										
170.0	RATE CHANGES	1,507						1,507		
180.0										
190.0	OVER STD REBLD	-643	-664			-500		-3,756	-6,484	-6,266
200.0										
210.0										
220.0	TOTAL VARIANCES	887	-1,328			-500		-4,408	-5,612	-5,444
230.0										

Figure 10.7

```
P&LRPT2     LIST

LISTING OF REPORT FILE P&LRPT2  (::::)

TITLES

HEADINGS
COL      1                   ------
                             ACTUAL
COL      2              LOS ANGELES
                             BUDGET
COL      3        ------
                           PRIOR YR
COL      4                   ------
                             ACTUAL
COL      5              --CHICAGO--
                             BUDGET
COL      6        ------
                           PRIOR YR
COL      7                   -----C
                             ACTUAL
COL      8              OUNCIL BLUF
                             BUDGET
COL      9        FS----
                           PRIOR YR
COL     10                   ------
                             ACTUAL
COL     11              -CHARLOTTE-
                             BUDGET
COL     12        ------
                           PRIOR YR
COL     13                   ------
                             ACTUAL
COL     14              - SYRACUSE--
                             BUDGET
COL     15        ------
                           PRIOR YR
```

```
COL     16                ------
                          ACTUAL
COL     17           CONSOLIDATE
                          BUDGET
COL     18       D-----
                        PRIOR YR
NON   STD   COLUMN   WIDTHS
COL     1  WIDTH  11  CHARS
COL     2  WIDTH  11  CHARS
COL     3  WIDTH  11  CHARS
COL     4  WIDTH  11  CHARS
COL     5  WIDTH  11  CHARS
COL     6  WIDTH  11  CHARS
COL     7  WIDTH  11  CHARS
COL     8  WIDTH  11  CHARS
COL     9  WIDTH  11  CHARS
COL    10  WIDTH  11  CHARS
COL    11  WIDTH  11  CHARS
COL    12  WIDTH  11  CHARS
COL    13  WIDTH  11  CHARS
COL    14  WIDTH  11  CHARS
COL    15  WIDTH  11  CHARS
COL    16  WIDTH  11  CHARS
COL    17  WIDTH  11  CHARS
COL    18  WIDTH  11  CHARS
SELECTED LINES
**************
-ALL LINES IN MODEL
SELECTED COLUMNS
****************
COL   1  TO  18  EDITING    ,Z
LINE NUMBERS PRINTED
PAGE NUMBERS SUPPRESSED
TITLES PRINTED ON ALL PAGES
SYSTEM CALCULATES COLS/PAGE
*** END OF REPORT LISTING ***
```

11

DDP +
DP +
DWP +
WP =
IP

11

DDP + DP + DWP + WP = IP

As mentioned in the beginning of the previous chapter, this chapter will look at another possible alternative to the traditional data processing life cycle for business applications.

One of the hottest buzz words today is word processing (WP). The next step beyond word processing will be joint DP/WP applications. Before looking at possible joint applications, we should look at the history of DP/WP, what they have in common and where they are going. Therefore, let us examine the picture as to yesterday, today and tomorrow.

YESTERDAY

Figure 11.1 Yesterday

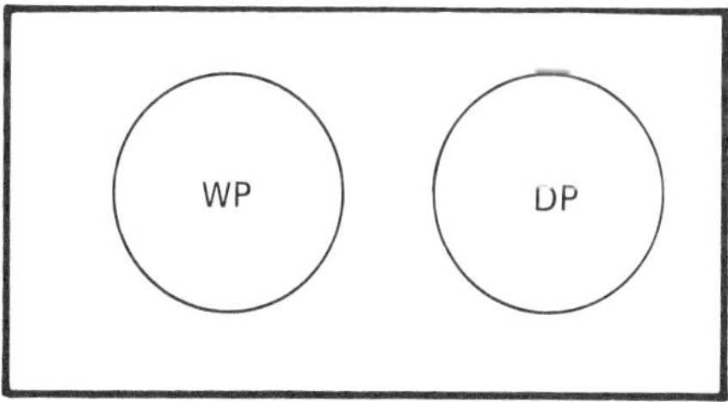

NOTE: This chapter is co-authored with Mona R. Bertrand who supervises the Word Processing Center at Carrier Corporation in Syracuse, New York. She has also written articles for several WP publications and speaks at conferences on word processing and management. The author is indebted to Mona.

Word Processing

The closest thing to WP, as we know it now, consisted of the all too familiar boss-secretary configuration. On a typical day, the secretary had the opportunity of "processing" approximately ten letters in final form, sandwiched between answering the telephone, making photocopies, bringing the boss and co-workers coffee, etc.

The human aspect at that time was ultimately highlighted and very much in evidence. Unfortunately, this arrangement did not promote efficiency nor did anything in the way of advancing technology or a secretary's method of getting work done. The electric typewriter was the secretary's only tool. The skills learned in high school or secretarial college were the only ones she was allowed to utilize, and the chances of getting a better position or updating her technology were nil.

The work load in an individual office peaked so that emergency deadlines found the secretary frustrated and indulging in copious amounts of correction fluid and correction tape and eraser crumbs jamming the typewriter, only to be met the following week by perhaps another emergency or, worse yet, a few days of very little to do. There was no leveling of work loads and no one to share responsibilities, only the grim determination to go it alone.

Automation in the office was unheard of. Factories and DP had undergone a dramatic change but the office remained essentially the same as it had been for 100 years, when the typewriter had been invented. The secretary remained the "office wife" and until the advent of WP, as we know it now, any deviation of that role was unlikely and unheard of.

Data Processing

Early 1960s. Most systems were standalone, noncompatible systems located at major operating units. Major applications were in the area of financial reporting. Mass storage devices began to be adopted widely, but magnetic tape was still the primary medium for storage of the data files.

Late 1960s. Most systems were becoming standalone compatible systems and the software was becoming standardized. New applications were in the area of financial control. There was a more extensive use of data communications. Mass storage devices were much larger, providing storage for billions of characters of data on line to the computer.

This was the era of decentralization, the advantages of which were as follows:

Improved effectiveness

Lighter economic consequences due to failure

Less training and bureaucracy

Less competition for service

Less sophistication needed

Lower communication costs

Early 1970s. During the early part of the decade we began to see the concentration of processing power at control sites with terminal access to large control data bases. The new applications were in the areas of order entry/inventory.

This is the era of centralization. The following were the advantages of centralization:

Lower overall cost of equipment and systems development cost

Fewer personnel problems and support personnel

Better DP cost control

More sophisticated systems possible

Ability to run large jobs

TODAY

Figure 11.2 Today

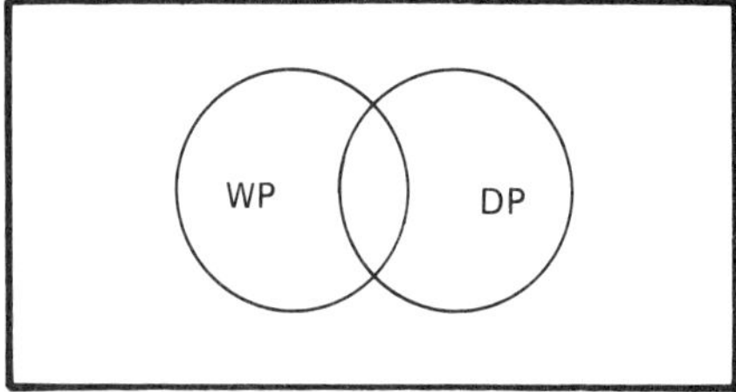

Word Processing

WP is separated from DP by only a thin line of technology. Though some WP centers are sponsored by the DP departments of large corporations, WP is still looked upon as unique in its stature and makeup.

Currently, first and foremost, WP is *people*. These people possess special talents, coupled with a spirit of adventure, and have ultimately conquered their fear of change along with subsequent acquisition of certain skills. WP personnel are accustomed to an atmosphere where nothing "stands still," and today's adequate station may be obsolete tomorrow. In a WP center, utilization of the latest and most modern, automated and computerized typing equipment, as well as efficient machine dictation, is a way of life.

Human resource requirements in this specialized field include a willingness to share technology and the patience to do so. One highly noticeable characteristic of most WP people is their full and complete knowledge of the systems with which they work, along with their patience and exceptional willingness to get the job done. In essence, they speak the specialized languages of the systems. In addition, as a matter of course they are expected to pitch in to teach others in the corporation as well as new center personnel the methods of gaining the most efficient productivity from this highly specialized equipment. As part of the daily routine, they fully employ the WP work sharing and team work concepts.

Contrary to popular belief, the WP function is not an isolated one within the corporate organization. There exists not only the normal amount of interpersonal relationships and day-to-day stresses and strains found in most other departments, but due to the overlapping scope of endeavors, WP people handle situations and cope with more varieties of personalities than most other departments ever do.

Yet, a great many more talents are required of the WP operator other than being capable of dealing effectively with people, handling sophisticated typing equipment and training others. There is an overriding requirement to select people with a flair for writing procedures and training manuals, and those with foresight and cognizance to make equipment evaluations. Moreover, superior editorial skills are required—that innate, anticipatory power to interpret vague references, to rearrange, to reemphasize, to restructure, to make clear.

Among the attributes of a good word processor is, as already mentioned, patience—the patience to deal with a highly motivated perfectionist, the individual who not only wants the work done on time, but also wants it to be letter-perfect.

The ability to select new equipment, evaluate it and actively participate in subsequent training programs is an inherent capability requirement of WP personnel. This *ability* is an ongoing process that never ends because the technology is constantly changing and being updated at an accelerated rate. Yet, mere acceptance, knowledge and evaluation are not enough. The ability to fit the new pieces of equipment to specific jobs is perhaps more important, the criteria always being, will it do the job, and if so, will the customers be satisfied with the output?

The human aspect is highlighted by the operators' pride in their unique skills, the advanced technological environment in which they work, as well as their ability to produce sizable volumes of work. The tools they have learned are highly sophisticated, enhancing their position within the company, as well as their own self-esteem and pride in their work. These newly acquired skills have opened up new career paths for yesterday's "mere secretary" who had only the *aspiration* to supervise and/or conduct training on a managerial level.

Work load sharing and leveling are two of the many advantages of WP. No longer is error correction a problem, nor is revision, extensive or otherwise. Responsibility for deadlines and priorities is shared by operators who can divide a lengthy job into as many parts as necessary, and yet complete, format and print it as one. Planning is done by a knowledgeable supervisor who knows the work load, the operators' abilities and the capabilities of the equipment.

Yes, indeed, today WP is people—people plus machines plus know-how. Basically, there is no comparison of today's WP concept to yesterday's boss-secretary configuration when you consider efficiency, production statistics, talent and employee opportunities.

Data Processing

We see the beginnings of distributed data processing (DDP) which means having processing power and data bases at remote locations. The new applications are transaction driven.

This is the era in which the pendulum is swinging from centralization to DDP. DDP is a cross between centralization and decentralization. With DDP, the computer power is decentralized at the remote sites to capture and process data at its source and transmit large amounts of clean data to the central site. Therefore, the computer power is decentralized but the control of the computer power is centralized. The concept of DDP has been around for a long time, but only recently it has been possible to achieve. This move was made possible by the development of inexpensive minicomputers and cheaper data communications.

A real DDP system is a network of powerful, self-sufficient satellite systems communicating with each other and/or the host computer. This is an environment where business information is controlled by the people using it. Data communications must be efficient and simple to add to existing data processing systems.

The following are the advantages of DDP:

User has local processing and data storage capability

User has integrated access to other computers

User controls his or her own data

Noncompetitive access to user

Reduces peak load demands on central computer power

Greater responsiveness to user

New applications can be quickly supported without costly additions to the central computer

Lower total system communications costs

Less training and bureaucracy

On the other side are the disadvantages of DDP:

Duplication of input, output and functions

Higher cost due to duplication of hardware, software, data, space and people

Application size and complexity restrictions

Possible incompatibilities

Restricted growth

More difficult management and control of operations, standards, application of development and data bases

Empire building

Limited equipment selection expertise

Limited vendor relations/negotiating power

TOMORROW

Figure 11.3 Tomorrow

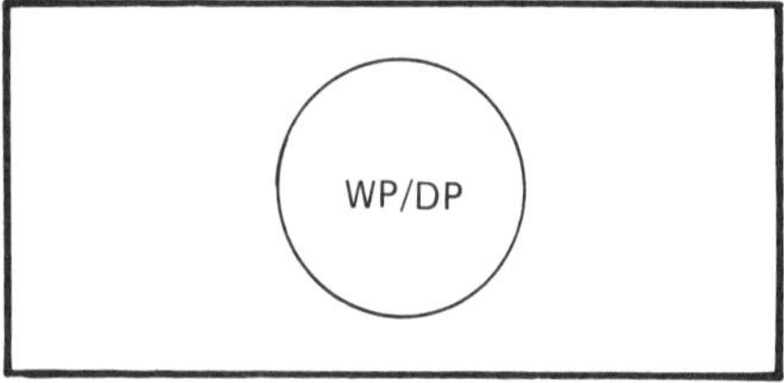

The future will bring broader use of DDP elements and redistribution of previously centralized applications. Computers and their networks will interface with other organizations. Therefore, we shall see an overall increase in computer usage with both new and expansion of existing application areas.

Because of their expertise, the DP people will help in the evaluation and selection of WP equipment. WP and DP people will be sharing the same work stations and same communication system. Also, some WP and DP data files will be compatible.

Some work stations will be a multifunction unit with various work-keys. At a touch, a work-key could be used for word processing, source data entry, central computer inquiry and standalone data processing. These multifunction units can function in work station clusters and communicate with other dispersed work station clusters and/or the host computer.

DDP + DP + DWP + WP = IP

Distributed DP plus DP plus distributed WP plus WP will be combined together for the next evolutionary jump into information processing (IP). Top management within each company will mandate this jump because they will realize the economics from sharing equipment, personnel and data.

The following common equipment features can be used for both DP and WP.

Hardware

The hardware will be storage devices, video displays and printers. The intelligence of the hardware can be any one of the following: hard wired, microprocessor based, processing capability in a host computer or a terminal with its own processing capability.

Software

WP software could be added to the resident DP software available which could include user-oriented application software packages that greatly expand the usefulness of the system.

Communications

Communication options can allow WP to transmit information to other dispersed work station clusters and/or the host computer. This information could become part of a data base to be used in DP applications and/or transmitted back to WP for their own use.

Functions

The major function could be a service bureau. The intermediate functions could be a charge-back system and/or production statistics system. The minor functions will be create, delete, list, merge, sort and update.

Applications

Some of the applications that can be used are accounts receivable-past due, long texts, marketing surveys or literature and mailing labels. Other applications would utilize the SORT function. For example, if we had 100 different catalogues on file by part number, we could select different catalogues and sort them down together by part number. Finally, we could print a new catalogue or report.

Some other uses are applications for in-plant phototypesetting and transmitting material electronically to remote locations via telecommunication modes. Most phototypeset material becomes copy for some form of duplicating or printing process. The material could be financial reports, annual summaries, booklets, pamphlets, catalogues, directories or advertisements. In this type of arrangement the phototypesetting machine itself becomes a peripheral to be controlled by the computer.

As time goes on, we shall find many more joint applications. Some of the above applications have programs which select certain variable data and output it to WP on storage devices. Later on the variable data will be blended with the fixed data that had previously been stored on storage devices by WP.

In conclusion, WP and DP will be parts of the total corporate information system.

12

The End Is Just Beginning

12

THE END IS JUST BEGINNING

Traditionally, we have looked first at the application the user needs and then backed into the hardware/software required to meet user requirements. Let us now take the *reverse* approach and look at only the hardware/software that is multifunctional and flexible to grow with.

We shall look at the integrated electronic office of the future which is available now. This super turnkey system *combines* data processing functions with the electronic counterpart of the typewriter, mail room and filing cabinet. The name of the company that sells this super turnkey system will be called Startpoint. Startpoint's system is comprised of the following functions: data processing, word processing, electronic mailbox and communications management.

DATA PROCESSING

Distributed data processing is carried to its highest level with Startpoint's EXPAND system. EXPAND has the following advantages.

Provides a computing network which can grow to almost any size or configuration.

Each user has easy access to all of the systems resources which are user defined. The resources include print spoolers, communication links and a common data base, regardless of location.

Easy to reconfigure the system. Processors and peripherals can be added, deleted, moved or functionally altered without disrupting the rest of the system.

The same equipment can be utilized for word processing, electronics mailbox and communications management.

EXPAND has the following components and facilities:

The three types of functional processors are used for applications, files and communications. The processors are intelligent video display keyboard terminals that are easy to use. Also, the system and user memory of each processor can be easily upgraded.

The communications processor is essentially a controller used for polling or addressing the line discipline for external communications.

Processors within one EXPAND system can participate with processors in other EXPAND systems.

The interprocessor bus attaches the processor together which makes high-speed data exchange possible and addresses each processor within EXPAND.

The direct channel interface allows a mainframe computer to participate with an application processor in an EXPAND system.

The three types of peripherals used are work station, disk storage and printer. The work stations are attached to the application processors.

A work station is a low-cost video display keyboard terminal. The work stations enable a business to share the resources of a single application processor economically. Each application processor can have up to 24 local or remote work stations so that users can enter data, execute programs, communicate and perform a wide range of other tasks.

The three types of keyboards available for the terminals are standard, data entry and multifunctional.

EXPAND has its own application development software called TOOLBOX. TOOLBOX uses Startpoint's programming language called TOOLBUS for application development. TOOLBOX will simplify system design, increase programmer productivity, simplify operator training/maintenance and reduce program development and system life-cycle costs. The TOOLBOX package includes skeleton programs, data dictionary, generators and a subroutine library.

The five programming languages available are BASIC, COBOL, FORTRAN, RPG and TOOLBUS.

See Figure 12.1 for a possible EXPAND configuration.

Figure 12.1 Possible EXPAND configuration

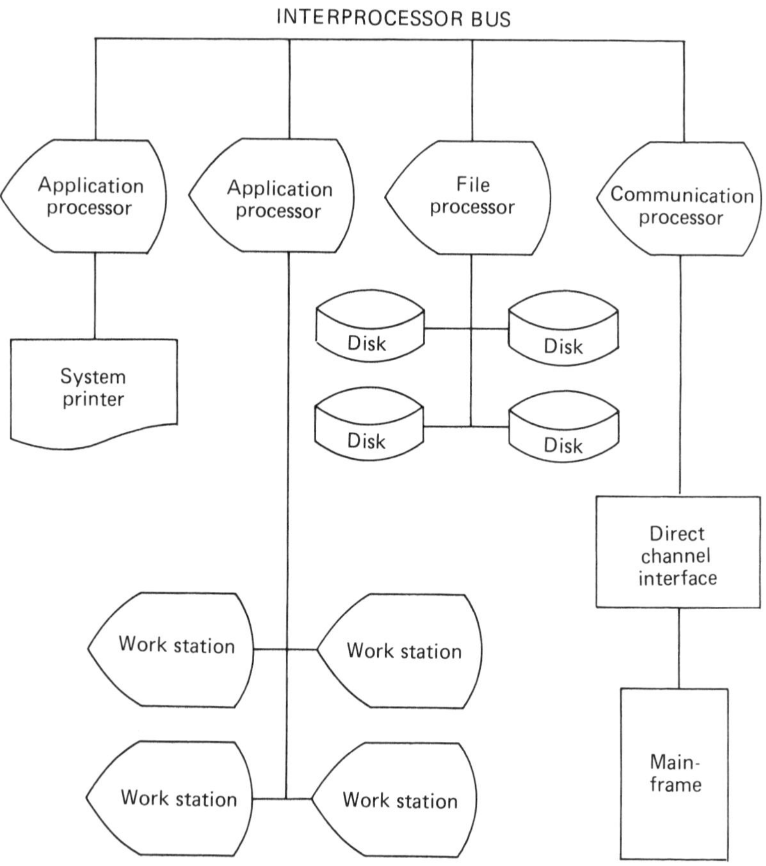

WORD PROCESSING

Word processing is just an extension of data processing. The word processing files cohabit with files for data processing and with files for communications management. Having those multiple files available at one location is a unique advantage.

A single work station can generate a letter discussing accounts receivable performance for ten regions and dig into the data base to get numbers for the letter or an attached report. Then instead of typing ten identical reports and cover letters for ten regional managers, the terminal can send them out electronically over telephone lines to printers or terminals in the branch offices. By having word processing on the same disk (memory) as communications management, we can queue a word processing message into the telephone system right along with voice messages.

Document management is composed of maintenance and retrieval commands. The following are the commands:

CREATE — edit a new document

MODIFY — edit previously created document

ADD — copy data processing document or file into word processing library

DELETE — remove document from document library

EXTRACT — copy document from document library into data processing environment

PACKUP — reorganize a document library to speed access and save space

DOS — return to disk operating system (date processing)

The document search and scan facility is made up of inspection and location commands. The following are the commands:

CATALOG—displays names of all documents in document library

LOCATE—displays names of documents in which a specific text segment occurs

SEARCH—displays any occurrence of a specific segment of text

The six edit keys on the keyboard are insert/delete, return, backspace, shift, tab and command.

Word processing's user-controlled format elements are as follows:

Page justification

Even, right, left, or center

Screen displays actual justification

Line spacing

Tab settings

Page headers

Page footers

Page numbering

Margins

Top, bottom, left or right

Since word processing has straightforward commands, one can:

Skip lines

Delete blocks of text

Boldface and insert left margin

Underline

Move text

Subscript

Insert one document into another

Superscript

Locate a word

Jump to a new page or phrase

Prompt the printer operator

Save blocks of text

Look at embedded format structure

Equate one key to a frequently used phrase or function

Insert blocks of text

Scrolling columns up and down from beginning to end and back

Scrolling columns from left to right from beginning to end and back

Printing is on high-speed printers and word processing letter-quality printers. Specific pages or the whole document may be printed or spooled for later printing.

ELECTRONIC MAILBOX

The electronic mailbox is just an extension of the word processing system and communications management which provides easy entry of message information as well as interactive reviewing and filing of electronic messages. This is a vehicle for high-speed message pickup, routing and delivery.

The electronic mailbox has the following advantages:

It is a timely and cost-effective communication system to make sure that the necessary information is being directed and moved to those who need it in an efficient and controlled manner.

It will reduce the need for separate facilities such as mail rooms, copiers and communication systems.

It will eliminate duplication of tasks such as retyping or addressing multiple envelopes, thereby increasing office productivity.

The system decides the best manner of distribution and ensures that proper priorities are employed.

The system has been designed to build upon existing equipment with minimum cost and effort.

The company can set the priorities for each message based upon importance and cost. The four priorities are:

OVERNIGHT—the message will be delivered and waiting for the recipient first thing in the morning

REGULAR—the majority of messages that will have same-day delivery

URGENT—the "urgent" messages are routed ahead of the "regular" messages for the same-day delivery

IMMEDIATE—the crisis message will go the moment it enters the system and be delivered in a few seconds.

For confidential correspondence, the system offers encryption which the sender orders by using a special code. The content of the message is scrambled while the message is in transit. The message will be decrypted if the recipient's definition file includes the special code used by the originator. Also, the system allows further security with verification of delivery and acknowledgement of receipt features.

Management and accounting information is available on a scheduled or on-request basis. The reports include information about the system's performance with charge-backs by user, terminal and/or department. Also, summary reports are provided.

COMMUNICATIONS MANAGEMENT

Communications management gives the user the tools needed to make trade-offs in telecommunications in terms of cost and service that other business areas have long possessed. These turnkey systems offer the attributes of being easy to install and simple to use.

Four communication systems are available. One, a long distance control system that reduces long distance costs an average of 15–40% through a combination of least-cost routing, queuing and call buffering. Management and accounting information is available on a scheduled or on-request basis. The reports include detailed and summary information on-line utilization on every call placed within the corporation by individuals, department and/or division. The electronic mailbox has the ability to work in conjunction with the long distance control system to transmit messages when there are no voice messages queue and during nonpeak calling hours. The long distance control system acts as a traffic manager by interleaving message traffic into the valleys between peak telephone traffic usage. Essentially, the message is given a free ride, since the user is already paying for unused flat-rate capacity. See Figure 12.2 for an overview of message and voice traffic usage.

Two, a short distance control system that is an extended application of the long distance control system for monitoring local and toll charge calls. This system helps the communications management to allocate costs accurately, adjust facilities and optimize service for local calling.

Figure 12.2 Message and voice traffic usage.

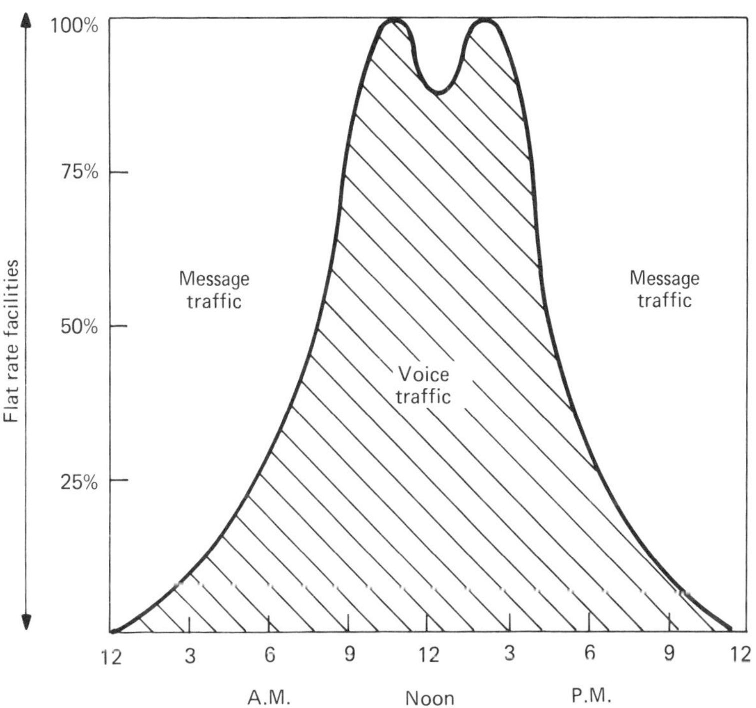

Three, a multilocation control system that is a combination of the long/short distance control systems for a multilocation company that needs a simple means of centralizing its communications. The multilocations take care of the details of actual call placement, but communications management and reporting occurs at the user's headquarters.

Four, an incoming-calls control system provides efficient processing of high-volume incoming telephone calls and makes optimum use of existing telephone facilities. Flexibility in call processing and queuing allows a company to operate more efficiently and economically while providing better service.

CONCLUSIONS

The integrated electronic office of the future moves computer technology into the office now with systems that look to the future as they meet current needs. The integrated approach means having a family of systems using general-purpose processors that can do a variety of jobs. As one can see, the end is just beginning.

INDEX